The story of
MADAME CURIE

By ALICE THORNE

Illustrated by Dan Dickas

SCHOLASTIC  BOOK SERVICES

Published by Scholastic Book Services, a division
of Scholastic Magazines, Inc., New York, N.Y.

Single copy price 50¢.
Quantity prices available on request.

3rd printing . . . . October 1963

Printed in the U.S.A.

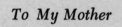
To My Mother

Grateful acknowledgment is made to Mme. Eve Curie Labouisse and to Doubleday & Co., publishers, for generously permitting the use of certain dialogue from the book MADAME CURIE, *by Eve Curie, since these were the words actually spoken at the time some of the events in the story happened. It is the author's hope that when the young readers of this book are older, they will want to read Eve Curie's own, beautiful story of her mother.*

CONTENTS

CHAPTER ONE

The Secret School

THE classroom was very quiet. Through the big windows on one side could be seen the leafless trees of the Saxony Garden, white now with the first snowfall. But not one pair of eyes strayed from the history books which twenty-five little girls were studying so earnestly.

It was not that they feared the teacher, Mademoiselle Tupalska, though she did have a plain face and a severe manner. On the contrary, "Tupsia," as they called her behind her back, was much admired by her pupils. For this was the year 1877, and the school was in Warsaw, Poland.

A large part of Poland had been conquered by

Russia. It was forbidden to teach Polish children the history of their own country or even their own language. But Tupsia was doing just that, although the Russians had spies everywhere in Warsaw.

A shaft of pale November sunlight crept along the rows of schoolgirls. They were all dressed alike in navy-blue serge with starched white collars. The sunbeams turned to gold the light hair of one little girl in the third row, and played among the curls that had escaped from her tight braid. But Marya Sklodovska, whose nickname was Manya, never noticed. She was deep in the book she was reading.

Suddenly there came the faint sound of a bell. With a start, Manya came back to the present. She listened fearfully. Was it the signal? Yes! Two long rings, two short rings.

Every head came up. Quick hands grabbed all the Polish history books off the desks and scooped up all the papers. Four girls ran along the rows holding out their aprons. The books and papers were tossed into the aprons, and the four scampered through a door leading to the boarding students' rooms.

The other girls swiftly took sewing materials

from their desks. They scattered thread, needles, and scissors about. With hands that shook a little, they began to embroider little squares of cloth.

Tupsia shot a last look along the rows of desks. Then she picked up a big book printed in Russian.

Just as the four girls who had hidden the Polish books returned to their seats, trying not to pant, the outer door opened.

There stood Inspector Hornberg, who had been put in charge by the Russians of the private schools of Warsaw. He was a bulky, heavy man dressed in a tight-fitting yellow and blue uniform. His close-cropped hair outlined a bullet-shaped head and a fat face. And when Manya saw his cold, steely eyes behind the thick glasses, she felt sick with fear and hatred.

With the inspector was Mademoiselle Sikorska, the directress of the school. Mademoiselle Sikorska was outwardly calm as she glanced at the teacher and the pupils. But inwardly she was terribly anxious. There had been so little time to warn the class of the inspector's arrival.

But the inspector found nothing inside the desks when he lifted a lid here and there as he walked along the rows. And the young hands that held the embroidery squares did not tremble now. The twenty-five girls sat quietly as Tupsia calmly invited Inspector Hornberg to take a chair.

"We have two sewing classes a week, Mr. Inspector," she explained. "I read to the children while they work."

"And what have you been reading to them this week, Mademoiselle?" Inspector Hornberg demanded.

Tupsia held up the book. "Russian fairy tales," she said.

The inspector gave a grunt of approval. "Now," he said, "I should like to question one of your pupils."

Manya's heart pounded, and she tried to seem smaller behind her desk.

"Please, please don't let it be me," she thought, in panic. But she knew it would be. Although Manya was only ten, two years younger than the other girls in the class, she was by far the best student, and she spoke Russian very well.

"Manya Sklodovska, please stand," Tupsia ordered quietly.

Manya rose from her seat without a word, but she was tense and trying not to tremble. She felt so hot that she wondered if her face had gone red.

"Recite the Lord's Prayer—in Russian," Inspector Hornberg snapped.

Manya recited the prayer without hesitation in a low voice, trying not to show her feelings.

"Now name the members of the Imperial Russian family," the inspector ordered.

"Her Majesty the Empress, His Imperial Highness the Czarevitch Alexander, His Imperial Highness the Grand Duke—"

"That will do," Hornberg interrupted. "Name my title."

"*Vysokorodye*," Manya answered.

The inspector puffed out his chest a little.

"Now tell me," he demanded, "who is our ruler?"

Manya's face went pale, and her deep-set gray eyes flashed angrily before she quickly lowered her eyelids. She opened her mouth, but the words would not come.

"So, my little Polish patriot, you do not wish to tell me who rules over us," the Russian inspector said, scowling. "Answer me!"

Manya swallowed hard, and at last she could no longer keep her voice from trembling. "His Majesty Alexander II, Czar of All the Russias," she said.

"That is more like it," Hornberg grunted, rising from his chair. "Now, Mademoiselle Sikorska, I wish to visit one of the other classes."

"Certainly, Mr. Inspector," Mademoiselle Sikorska agreed and led the way to the door onto the landing. The inspector followed without a backward glance. When the door had closed upon them, a sigh of relief ran along the rows of frozen school-girls. They began to stir again.

Mademoiselle Tupalska looked over at Manya, who had returned to her seat and now sat limply behind her desk. "Come here, Manya," she said gently.

Manya stood up and walked to the teacher's desk. Tupsia's homely face was full of pride and pity. Without a word, she put her arms around the little girl and kissed her.

Manya burst into tears.

CHAPTER TWO

The Tower of Chairs

WHEN school was over for the day, Manya and her sister Hela ran to the cloakroom for their heavy coats and mufflers. Hela was two years older than Manya and very pretty.

"You did well today, Manya," Hela said, winding a bright red muffler over her shining hair. "For a moment, though, I was afraid you were going to refuse to answer the inspector."

"I would not dare refuse," Manya said bitterly, stamping her feet into her boots. "But how I hate his fat, pig face!"

"Are you ready? Come along then," Hela said gaily. "We mustn't keep Aunt Lucia waiting."

Madame Sklodovska, the children's beautiful mother, was very ill, and loyal Aunt Lucia was in the habit of calling for them at the school. She was standing in the snow at the foot of the steps now, waiting. Hela ran to her, full of excited chatter about the inspector's visit. Manya followed more slowly, still upset and unhappy.

Aunt Lucia glanced at the silent little girl and said cheerfully, "How would you children like to go down to the Vistula and help me pick out apples for the winter?"

"Oh, good! Good!" Hela cried happily. "May I pick out the apples, Aunt Lucia? May I?"

"And I too?" Manya asked timidly. Her gray eyes were shining now.

"Of course, you may both pick them out," Aunt Lucia said. "Now, let us walk fast. It is bitter cold today. And besides, we shall want to stop at the Chapel and say a prayer for your dear sister Zozia."

Zozia, the oldest sister, had died two years before. Manya, who was the baby of the family, thought no one in the world could ever tell such wonderful stories as Zozia had. The little girl missed her keenly, even though she still had Hela, her brother Joseph, and her favorite sister, Bronya.

Walking rapidly, with their schoolbags swinging from their shoulders, the two sisters and their aunt crossed the Saxony Garden. They entered the old section of Warsaw.

Here the snow covered high, peaked roofs and tufted the elaborately carved gray-stone fronts with white. They came to the Chapel of Our Lady and climbed the old, uneven red-stone steps.

Inside the church, Manya knelt near her sister and her aunt and prayed for Zozia's soul. Then she added another prayer, to ask God not to let her mother die. But in her heart, Manya knew that this was one prayer God might not grant.

Out in the cold afternoon air again, the three

picked their way carefully down the slippery steps that led to the Vistula River.

"I see them!" Hela exclaimed joyously, pointing to two long, narrow boats on the swirling yellow water. The boats were loaded with great piles of shining red apples.

"All right, children, go and tell the master of the boat on this side of the pier that you wish to pick out your own apples," Aunt Lucia said. "And be very careful to select the firmest ones."

Manya and Hela skipped over to the man who stood by the nearer boat. He was swinging his arms against the cold even though he was bundled up warmly. He gave the girls a large basket, and they went to work.

"If you find a rotten apple, throw it in the river," he told them cheerfully.

The children soon filled the basket with fine red apples. They chose one each to eat on the way home. Then Aunt Lucia paid the boatman and handed the brimming basket to a ragged boy who was standing near by. She gave him a coin and told him to take the basket to the Sklodovski apartment on Carmelite Street.

When Aunt Lucia and her nieces reached home, the boy had come and gone. It was five o'clock, time for tea. The dining room was crowded, for besides Professor Sklodovski and his family, there were ten boy boarders.

Manya sighed, remembering the peaceful, happy days when they had lived in a large, quiet apartment with no boarders. But that was before her father had had trouble with the director of the Russian school where Professor Sklodovski taught physics. Now, his salary had been lowered, and he was forced to take in boarding pupils to make enough money to care for his family.

Manya and Hela took off their snow-crusted boots and hung up their coats and mufflers in the hall closet. Then Hela joined the others around the steaming tea urn. But Manya first tiptoed along the hall to her mother's room. She opened the door carefully.

"Mamma is sleeping. I must not wake her," she thought, peering over at her mother, who was lying in bed with her eyes closed. Cautiously, so the door would not squeak, Manya closed it and tiptoed back down the hall. When she reached the dining room she made her way around the long mahogany table to where her father sat.

"Ah, my little Manya," the professor said, his eyes lighting as they rested on his favorite child. "You have color in your cheeks today, little rascal. I am glad. You have been too pale lately."

Manya pressed her cold cheek against her father's as she hugged him.

"We went to the Vistula for apples, Father," she said. "It was very cold by the river, but the apples were beautiful."

"Come, sit here by me and have some hot tea," he said. "You will soon be warm."

Professor Sklodovski was short and stocky. His clothes were always in order, and his speech was careful and precise. He had a short gray beard, and his face always seemed rather solemn and severe. But, although he often looked stern, he had a tender, loving heart.

When everyone had finished tea, the professor stood up.

"Time for study now, boys," he said.

The boy pupils piled out of the dining room and drifted off to their own rooms. Soon they could be heard mumbling, muttering, or shouting their lessons.

Meanwhile, a servant had cleared the tea things off the big mahogany table. And Professor Sklodovski and his children sat down to study also. Manya took her place beside her sister Bronya, who was three years older. Joseph, their fourteen-year-old brother, came in rather breathless. He had missed tea, but he knew his father would be distressed if he were late for his lessons.

Hela, who was sitting next to Aunt Lucia's daughter, Henrietta, whispered something in Henrietta's ear. Henrietta giggled. She nudged Bronya, who was sitting next to her, and in turn whispered in Bronya's ear. Bronya eyed Manya, who had already started reading her book. Bronya opened her mouth to form the word, "Later."

Hela nodded, her eyes gleaming mischievously, and opened her own book. She began to chant her Latin verbs in a strong, ringing voice. Without taking her eyes off the page, Manya put her thumbs over her ears so that Hela's voice would not disturb her. But soon Manya's thumbs came down.

For she was so deep in what she was reading that she no longer even heard Hela.

"Now!" Hela exclaimed in a loud whisper.

The three girls got up from the table quietly. They began to pile up chairs, one on top of the other. First they placed a red velvet upholstered chair behind Manya. Then they chose a straight chair to stand on top of it.

Professor Sklodovski had left the room a moment before, and Joseph merely watched grinning. Soon there was a tall, shaky tower of chairs looming up behind Manya. One movement on her part, and the whole thing would come toppling over.

"Will she never look up?" Hela whispered impatiently, when nearly half an hour had passed. Finally Manya finished the chapter she had been reading and straightened in her chair. CRASH! Down came the chairs in a series of loud thumps and bumps.

Hela, Bronya, and Henrietta howled with glee. Manya looked around, dazed, at the chairs which had skidded all over the room. Then she picked up her book and rose from the table. Hela ran behind a big, overturned armchair, in case Manya should chase her and tickle her.

But Manya was not angry. She looked at Hela and the others and said calmly, "That's stupid."

The older girls felt let down. The joke, it seemed, had fallen rather flat.

Manya already had forgotten the trick the other girls had played on her. For she had wandered over to the corner of the dining room where her father kept his physics apparatus.

In a shining glass case were small scales, glass tubes, small rocks and samples of metal, and a gold-leaf electroscope. The professor had used these things in his physics classes to show how the laws of nature work.

But now the Russian Government had forbidden him to teach his students how to use the instruments. So the physics apparatus remained at home in the glass case. It was an unfailing source of interest for Manya. She spent many hours standing there, as she did now, gazing into the glass case as though it were a crystall ball in which her future might be read.

CHAPTER THREE

A Gold Medal

Bronya, my papers, have you seen them?"

"Bronya, what shall I wear to dancing school to-night?"

"Please, Bronya, will you tie my bow for me?"

Bronya, in long skirts now, and with her golden hair high on her head, laughed as Joseph, Hela, and Manya all spoke at once. She tied the bow on Manya's dark red dress. Then she picked up the sheaf of papers that were right under Joseph's nose and handed them to him.

"Sit down now, all of you, and have your breakfast," she said. "It is after eight o'clock."

As they took their places around the table, Bronya ran a motherly eye over each one. Joseph was eighteen now, and wore the uniform of the University of Warsaw, where he was studying to be a doctor. Hela was still dressed in the navy-blue uniform of Mademoiselle Sikorska's school. But Manya, the youngest, already wore the dark red uniform of the Russian Gymnasium, or high school.

"Since you have only one party dress," Bronya said to Hela, laughing, "you really have no problem about what to wear tonight."

Although the professor still taught boy boarders, there was no money to spare. But the apartment in Leschen Street, where they had moved since the mother's death, was large enough for them to have their own rooms, away from the boarders.

"Pass the butter, please," Joseph asked Manya. "I can't be late today. We are going to have a test this morning."

"I don't see why girls can't go to the university," Manya grumbled, passing the butter. "Look at Bronya. She graduated from the Gymnasium last year with the top honors. She's the brightest one of us all. And *she* wants to be a doctor, too."

"Never mind, Manya," Bronya said, smiling. "Perhaps my chance will come someday. Besides, who would look after all of you and Father? You know what terrible housekeepers we have had."

"But Manya's right," Hela said, unexpectedly serious. "It isn't fair to you, Bronya."

"No," Joseph chimed in. "With your brains, Bronya, I shouldn't be surprised if you could become a student at the Sorbonne University in Paris."

Bronya sighed, but her eyes were glowing. "That is where I would love to go," she admitted. "They allow both men and women to study there. And I could get the finest education in the world there. Well—" she stood up briskly, "we don't have the money, and that's that. Finish your breakfast while I go and tie up the lunches."

Manya, who had finished, followed her sister to the kitchen. "I'll help you," she offered, and went to get the *serdelki*, delicious Polish sausage which she and Hela would eat at the eleven o'clock recess.

"I've already made the bread and butter sandwiches," Bronya said. "Just put in the apples, dear, and tie up the bags."

Manya finished packing the two lunches and swept up some bread crumbs. These she threw down to the pigeons in the courtyard.

"It's a beautiful day," she exclaimed joyously. "I'd better hurry, or Kazia will go on without me."

It was half past eight by now, and Joseph already had left for the university. Hela rushed into the kitchen, picked up her lunch bag, and departed for school.

"Good-by, Bronya," Manya called, as she too stowed her lunch in her schoolbag and started down the stairs. Two or three of the boy boarders were straggling in as she left the house.

"Poor Father," Manya thought, as she started off to meet her friend Kazia. "He works so hard."

But it was impossible to feel sad on such a May morning as this. Manya almost skipped along Leschen Street. She reached a courtyard in the entrance of which stood a battered bronze lion. She looked at the ring in the lion's nose. It was turned down. That meant Kazia had not yet come out.

"I'll wait two minutes," Manya thought, "and then I'll have to go on if she doesn't come."

Manya was about to turn up the ring in the lion's nose, to show Kazia that she had left, when her friend appeared.

Kazia Przyborovska and Manya had been best friends for nearly two years now. Kazia was not as good a student as Manya, who really loved school. But both girls enjoyed making fun of their Russian teachers. And neither girl had felt sad when the Russian Czar Alexander II was unexpectedly assassinated.

"Mother wants you to come to tea this afternoon," Kazia told Manya as, arm in arm, the two girls walked along the narrow street. "She says she will make you chocolate ice—you know you love it."

"And you must come to our house tonight, Kazia," Manya said. "There is going to be dancing, and even if they won't let us dance until we're fifteen, it will be fun to watch."

"I'd love to," Kazia agreed, as they entered Saxony Square. "I do wish they would let us dance now, though. We know all the steps, and—"

"Ooh! Kazia, we forgot to spit on the statue!" Manya cried.

"That's right, we'll have to go back," Kazia exclaimed, and the two girls ran back to a monument they had passed a few moments before.

Inscribed on the monument were the words: "To the Poles Faithful to Their Sovereign." Since this really meant the Poles who had been traitors to their own country and were faithful to the Russian ruler, Manya and Kazia almost never failed to spit on the hated statue when they went by it.

"We'll have to run now, or we'll be late," Manya said.

They reached the Gymnasium just in time. Groups of girls stood talking and laughing in the archway of the big, three-story building—German girls, Russian, and Polish. They greeted the newcomers gaily, and began to go into the high school.

"But Leonie!" Manya exclaimed, as she caught

up with another of her school friends. "Whatever is the matter?"

Leonie Kunicka's blue eyes were red with crying, and she looked wan and tired, as though she had not slept all night.

"Yes, Leonie," Kazia echoed. "What has happened?"

"It's my brother," Leonie whispered, starting to cry all over again. "He—he was plotting against the Russian Government, and someone betrayed him. The Russian police arrested him. And now—they are going to hang him at dawn."

Manya and Kazia turned white. They put their arms around their sobbing schoolmate and said nothing, for there were no words of comfort they could give her.

A harsh bell clanged. School had started. Manya looked at Kazia over Leonie's bent head and said quietly, "There will be no dancing for us tonight."

Kazia nodded agreement. And that night Manya and her sisters, and Kazia and her sister spent the dreadful hours before dawn with their heartbroken friend.

The memory of that dawn did not make life in a

Russian school any more pleasant for Manya. But the months went by, and at last, in June of 1883, the day of graduation came.

The auditorium was stiflingly hot. A heavy scent of roses hung in the air. There was a stir in the audience as Monsieur Apushkin, head of the high schools in Russian Poland, walked onto the platform to read the list of awards.

Professor Sklodovski, who was sitting in the audience with Bronya, Hela, and Joseph, leaned forward eagerly. Manya had been a brilliant student and was sure to graduate with honors.

The professor was not disappointed. A fanfare of trumpets sounded, and Monsieur Apushkin then announced that the Gold Medal had been won by Mademoiselle Manya Sklodovska!

Manya, wearing her good black dress, to which she had pinned a little bunch of roses, left her place among the graduating class and walked out onto the platform. She accepted the medal and shook hands with Monsieur Apushkin. Then she returned to her seat amid applause.

There were more congratulations, speeches, prizes. But for Professor Sklodovski, that was the high point of the exercises. Beaming with pride, he met Manya when the ceremonies were all over.

"You have done excellent work, Manya," he told her as they started for home. "You would have made your dear mother very happy, as you have made me happy."

Manya tucked her arm in his affectionately. "It is you who should take the credit, Father," she said. "I did not learn much from those Russian teachers. It was our wonderful Saturday night sessions with you that have taught me what I know."

For years, Professor Sklodovski and his children had gathered around the big tea urn one night a week, reading, talking, and studying different subjects. Since the professor knew a great deal about a great many things, there was truth in what Manya said.

"Now, Manya," her father told her triumphantly, "I have a surprise for you."

"What is it, Father?" Manya demanded. "Tell me."

"It is all arranged," he said. "You have studied hard for a long time. Now you are going to take a whole year off—just to enjoy yourself."

"Father!" Manya exclaimed delightedly. "How wonderful!" Then she shook her head, frowning. "But I can't do that. You have worked so hard. I have my high school diploma now. I can earn some money giving lessons to children."

"No, my child," the professor said. "You are not yet sixteen. You deserve to have some fun." He sighed. "I feel very sad that I have not been able to make enough money so that none of my daughters would have to work for a living."

Then he brightened and added with a little chuckle, "At least, we have no lack of relatives. And they all want you to visit them, Manya. It will do you good to get away from Warsaw for a while and have a vacation in the country."

CHAPTER FOUR

The Floating University

BOTH Manya's mother and her father had come from noble families who owned great estates. Many of their relatives still lived on the old farms. It was decided that Manya would spend part of her vacation visiting an uncle who lived in Galicia, in the shadow of the beautiful Carpathian Mountains. This uncle had three lively daughters about Manya's age.

Manya's arrival at her uncle's roomy house was all the excuse that was needed for a big party. Her jolly aunt and uncle joined their three daughters in making her feel welcome.

Some good-looking boys from nearby Cracow were invited to the house, and they soon found out that the visitor from Warsaw was a very good dancer. The party lasted until late in the evening. When it was over, Manya discovered that she had danced right through her shoes.

"It was wonderful," she sighed happily, as she went up the stairs in her stocking feet.

Her cousins laughed. "This party was nothing," one of them said. "Wait until we take you to the *kulig*."

"The *kulig*?" Manya repeated. "What is that?"

"I won't tell you," the cousin said teasingly. "You just wait and see."

One morning about two weeks later, the same cousin came into Manya's room and announced excitedly, "We're to go to the *kulig* tomorrow night,

morning when Manya danced the White Mazurka
—the last dance of the *kulig*.

After such a holiday as this, it was hard when the
time came for Manya to return to Warsaw. But her
father at last had given up the boy boarders. And
although he had had to move to a smaller apartment, Manya liked it better. It was quiet, and there
were no strangers running in and out.

"But I must start earning some money," she
thought, when she saw how tired her father looked.
So Manya put an advertisement in the newspaper,
offering to tutor children in arithmetic, reading, and
Latin.

One morning in November, a letter came direct-
ing Mademoiselle Manya Sklodovska to call at a
house across the city.

"Bronya!" Manya cried, when she had read the
letter. "I have an answer to my advertisement! Perhaps I shall get the position."

"It is snowing hard, Manya," Bronya said, coming into the dining room. "Be sure to bundle up
well."

"I will, Bronya." Manya already was in her coat

and boots. "I'll tell you all about it when I get home."

She started out gaily, although the snow-laden wind stung her cheeks. To save carfare, she decided to walk across town. The air was biting cold, but Manya was so excited that she scarcely felt it.

When she finally reached her destination and was admitted, the servant asked her to take a seat in the hall. While she was waiting to be interviewed by the mistress of the house, Manya looked about her. It was drafty in the hall and rather dismal.

"But perhaps the children will be sweet and clever," Manya thought. "Anyway, one should never let appearances bother one."

After about half an hour, a rather peevish-looking woman came down the stairs. With her was her son, a red-faced little boy of seven.

"Mademoiselle Sklodovska?" the woman greeted her. "This is my son Henryk. I wish him to have special instruction in reading. He is an extremely clever child but lazy. So far, he cannot read the simplest sentence."

"Perhaps if we work together, Henryk," Manya

said, smiling at the little boy, "it will become easier."

"N-a-a, it won't," snarled Henryk. "I don't want to know how to read anyway. I'd rather play ball!"

"Henryk!" his mother snapped. "Do not address the young lady so. Come, Mademoiselle Sklodovska, I will show you where you and Henryk may work. By the way," she paused, "how much do you charge for lessons, Mademoiselle?"

"Half a ruble an hour?" Manya suggested timidly.

"Half a ruble!" the woman exclaimed. "That is dreadfully expensive. However, no one else will work with Henryk, so we shall have to see what you can do. Are you willing to start now?"

"Yes, madame," Manya agreed.

But Henryk was not willing to start. With a howl of pure rage, he raced across the hall and up the stairs. Manya could see that there would be no lesson that day.

Later, there were lessons—if Henryk could be cornered in time. As the months passed Manya taught other children, too. And there were endless waits in other drafty halls, while little girls or boys lagged over their breakfasts. Sometimes, when it was the day Manya was to be paid, a flighty mother would laugh and say, "Oh, Mademoiselle Sklodovska, I am so sorry. I quite forgot to ask my husband for your four rubles. Never mind, my dear, I shall surely pay you when you come next week."

These would have been dreary days indeed for Manya, if she and Bronya had not met a young woman named Mademoiselle Piasecka. Mademoiselle Piasecka was a high school teacher by day. But at night she held secret classes in her home. Polish girls were forbidden to continue their studies after they had graduated from the Russian high schools. So great care was taken by them and by their teachers not to attract the attention of the Russian police.

Manya and Bronya joined the classes in natural history and other subjects. Because everyone feared the Russian police, it was decided to hold these classes in a different home each time, to avoid suspicion. Soon the secret students were calling their classes "The Floating University."

One night Manya and Bronya slipped out of the house where the Floating University had been held that evening. No one was with the two girls, as it was thought dangerous for large groups to be seen coming out. They walked quietly along the dark, narrow street toward home.

Bronya sighed. "I don't know what I should do without the Floating University," she told Manya. "How I wish I could go to the Sorbonne in Paris and study to be a doctor! I try and try so to save my money, but I haven't nearly enough."

Manya took a deep breath. She had been wondering for a long time how she could help Bronya. Now, she believed she had found a way—if she could persuade Bronya to agree.

"Bronya, listen," she began. "I know how we can manage it so you can go to Paris. Take the money you *have* saved, and go to Paris right away. I will work and send you the money I earn so you can stay on. Then when you have become a doctor, you can send me money enough to go to the Sorbonne."

"Manya, you would do that for me!" Bronya exclaimed. "But it isn't fair. You should have your chance, too. Besides," she added practically, "you hardly make enough money now for yourself."

"I know that," Manya went on firmly. "But I have

decided to take a position as a governess, some-where in the country. That kind of work pays well, and one also gets board and room free. I shall be able to send you lots of money!"

"Manya dear," Bronya said tenderly, "you would go far away from Faher, you would give up the Floating University—to do that for me! No, I can't let you."

"Oh, don't be silly," Manya exclaimed impa-tiently. "We're getting nowhere this way. You're twenty already, and you're wasting time. If you go now, in five years you'll be out of the university, and then you will help me."

"Five years!" Bronya exclaimed. "Why, it would be as though you had been banished to Siberia."

"Nonsense!" said Manya. "I'm seventeen now. I'll be only twenty-two by then. And there's so much we both want to learn. It's all there at the Sorbonne waiting for us. Please, Bronya, say you will do it my way."

"All right," Bronya agreed gratefully. "I will."

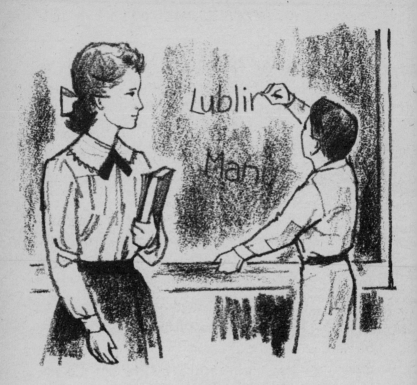

CHAPTER FIVE

A Letter from Bronya

MANYA gazed out her bedroom window and sighed. As far as the eye could see stretched the beet fields. Not a hill, not a tree broke the long, flat line of brown plains. The only upright objects to be seen were the dirty red-brick factory with its belching chimney, and the huts of the peasants who worked in the factory.

This had been Manya's view for a year now—ever since she had been hired as a governess to the children of the wealthy man who ran the sugar-beet farm.

"I never want to see another beet as long as I live," Manya thought, turning away from the window. It was time to awaken Andzia, who was ten and who loved to sleep late in the morning.

Manya took a quick look in the long mirror, to make sure her smooth, pale hair was tidy and her black dress orderly. She was nineteen now, and had grown slender and attractive. She herself did not realize this. She was too busy and too homesick to think of her appearance. Warsaw and Father were sixty long, lonely miles away. And there were seven boys and girls in the family here. Three were her particular pupils.

But Casimir, one of the older sons, realized how pretty the young governess was. When he came home on his Christmas vacation from the University of Warsaw, Casimir decided that Mademoiselle Manya was the loveliest girl he had ever met. And he fell in love with her. Manya was fond of him, too, but her work with the younger children kept her busy from morning to night.

She closed the door of her room now, and with quick steps went down the hall to Andzia's room.

The little girl was cuddled up under a great down-filled quilt, sleeping soundly. She was a pretty child. While she was asleep her face did not show the spoiled, cross expression it would have when she awakened.

Manya crossed the large room and opened the heavy curtains to let in the cold winter sunlight. She adjusted the draft of the huge porcelain stove that stood in one corner. Then she returned to the big bed and spoke to Andzia gently. The girl stirred and mumbled under the quilt but did not rouse.

"Come, Andzia, it is time to get up," Manya said again, and shook the sleeper's shoulder a little.

Andzia opened her brown eyes and closed them again. "Please, dear Mademoiselle Manya, go away and let me sleep," she said drowsily, and turned her face away.

Manya laughed and pulled the quilt away from

Andzia's face. "You must get up now, Andzia," she said firmly. "We have a great deal of work to do today."

Andzia opened her eyes again, and now her mouth held its usual pout.

"Why must I always get up so early?" she complained. "I *hate* getting up in the morning."

"I know you do," Manya observed, smiling, "but we can't always do just as we want."

Andzia was wide awake now, and had decided to be stubborn. "Today," she announced, "I am *not* going to get up."

"Yes, you are, Andzia," Manya said, more firmly this time. "And right away."

"I won't," Andzia cried. She was working herself up into a fine temper. "I *won't!*"

Suddenly all the boring days Manya had passed here, the dull people who surrounded her, the discouragement she felt about her studies, came to a boil within her. She could not leave this place until Bronya finished the four remaining years of her training in Paris. Bronya needed every kopeck of the twenty rubles Manya sent each month.

"I'll put up with this place as long as I have to,"

Manya thought. "But I'll not endure any more scenes and tempers from Andzia."

Keeping outwardly calm, Manya pulled back the quilt and blankets and seized Andzia by the wrist. The girl squealed with rage and surprise. Manya was graceful and delicate looking, but she was surprisingly strong. Without hurting Andzia in the least, she pulled her to a sitting position and right out of the bed.

Andzia glared at her governess but soon dropped her eyes before the deep, intense gaze of Manya's gray eyes.

"Now, mademoiselle, we shall have no more of this nonsense—ever," Manya told her. "Go and wash your hands and face, and then I will help you dress. You may wear your pretty green wool, if you wish."

Andzia's sulky face brightened, and she decided not to scream after all. The green was her favorite dress—as Manya knew.

"Yes, Mademoiselle Manya," the child said meekly. Suddenly she flung her arms around Manya. "I am naughty, mademoiselle," she exclaimed. "But I do love you."

"I know you do, Andzia," Manya said kindly. "Come now, let us hurry. It is nearly time for breakfast."

Andzia skipped over to the big washstand, and Manya followed to pour water into the bowl from the large blue pitcher. The little girl hummed merrily, her temper quite forgotten. Manya helped her to dress and talked cheerfully. But within she was upset and weary. At home there had been no scenes and tempers, and she would never get used to them.

After breakfast, Manya taught Andzia reading and spelling and arithmetic. It was slow work, for Andzia did not have a good memory, and also there were several interruptions. After each one, the little girl found some excuse to delay the return to work. By one o'clock Manya was quite as eager as her pupil to stop for luncheon.

In the afternoon, though, it was different. Now Manya's pupil was the older sister, Bronka. She was far more intelligent than her younger sisters and brothers. She and Manya had become good friends the moment they met.

"Bronka, I have an idea," Manya said today, as they sat in a wide window seat of the high-ceilinged study. "You know, the children of the peasants who work in the beet fields can neither read nor write, most of them. Why don't we start a class for them? We could have it in my room. It is large, and there is a separate entrance from the courtyard. We wouldn't be disturbing anyone."

She pushed back the blue velvet curtain and gazed out the window.

"See those children out there?" she went on. "They want to learn, but it is forbidden to teach them how to read and write in Polish. There are no schools for them, and scarcely any books around here. But we could help them."

"Manya, I think that's a wonderful idea!" Bronka exclaimed. Her big brown eyes were dancing with eagerness. "You know about so many things, and I will be your assistant. When can we start?"

Manya frowned. "We'd better ask your father's permission first," she said thoughtfully. "It is true we are far north of Warsaw here, but the Russian spies are everywhere. If we were caught teaching the children of peasants, we would be punished severely."

"I don't care," Bronka said recklessly. "Don't ask my father. Let's take the risk. It would be so much fun—just like playing school."

But Manya, remembering the dangerous days of the Floating University, insisted on asking permission of the master of the house. He granted it. So Manya and Bronka went to the peasants' huts and told the children's parents about their scheme.

Soon ten children had promised to come to the outside entrance of Manya's room at five o'clock. Manya and Bronka hurried home to make ready for the class. They went up to Manya's room and placed chairs in rows where the light was best.

"We'll need lamps by five o'clock," Manya remarked. "It gets dark so early these days."

"I'll have the servants bring some lamps," Bronka offered. "And I'll go down to Father's study and get some paper and pencils. I'll be right back."

At five o'clock all was in readiness. And soon Manya and Bronka heard footsteps on the outside stairs. They opened the door, and in filed ten solemn little peasants.

"Come in, children," Manya said, as some of the

younger ones hung back. "Take any seats you like at the table."

Manya and Bronka had drawn a long pine table away from the wall, so the children would have something on which to place their pads and pencils.

"Sit down," ordered one sturdy little boy in a loud whisper. "The lady says to sit down."

A flaxen-haired little girl of seven crowded close to the boy's side.

"Let me sit by you, Lublin," she whispered. Evidently Lublin was the leader of the little group.

After a good deal of shuffling and commotion, the children settled down at the long table and fixed their bright eyes on Manya.

"Can any of you write your own name in Polish?" she asked them.

The children shook their heads.

"Then that's what we'll learn to do first," Manya told them.

It took a long time and a good deal of patience on Manya's and Bronka's part. But the day finally came when Lublin and some of the other children strutted over to the blackboard the girls had hung on the wall. Carefully the children wrote their names in large letters.

"I am very proud of you," Manya said. Indeed, it made her happy to see how pleased they were with their accomplishment.

Poor Manya—there was not much these days to be happy about. For something had happened to make her life in this house almost unbearable. Young Casimir had told his parents that he was in love with the charming young governess and that he wanted to marry her.

This announcement created an uproar. Casimir was informed that no one in his family would be allowed to marry a governess. And although Manya was still treated politely, the family was no longer friendly.

Manya longed to go home to Warsaw, but she could not. Bronya still needed her help. So for three long years she taught the children of the man who ran the sugar-beet farm. Then she received good news from her father.

The professor had become the head of a school in a town near Warsaw. At last he could send money to Bronya, and Manya could come home.

After the years of exile in the beet country, Warsaw looked like heaven to Manya. She was lucky enough to find another position as governess, this time with a Warsaw family. That meant she could see her father every day. And she rejoined the Floating University.

Manya had kept up her studies in physics and chemistry with what few books she had been able to find at the beet factory. She had learned to work alone. But it had been slow and difficult. Now, she found that in a building in Warsaw, the Floating University had a hidden laboratory.

This was the first laboratory Manya had ever seen. All the wonder and excitement she had felt as a child, standing before her father's physics apparatus, came back. She could hardly wait for nights and Sundays. Then she would rush to the laboratory and try to perform the experiments about which she had read. And she dreamed—without much hope—of someday studying science at the Sorbonne University in Paris.

It looked for a while as though this hope would remain only a dream, for Manya had very little money saved. And she knew it would cost several hundred rubles to attend the university, besides paying for her room and board.

But one morning the postman delivered a letter that was to change Manya's whole life. The letter was from Paris—from Bronya.

Manya read the letter through, and then she read it again. Her heart began to pound, for Bronya was asking her to come to Paris! Bronya wrote that she was about to be married. She wanted Manya to spend a year with her and her husband in their new home—and attend the Sorbonne!

CHAPTER SIX

The Great Adventure

MANYA looked up the long line of tracks at the railroad station in Warsaw.

"Father dear," she said, "I shall not be away

long. As soon as I have finished my studies, I will come back and we will never be separated again."

"I shall miss you, Manya," the professor sighed, "but you must work hard and profit by this opportunity."

It was over a year and a half since Bronya's exciting letter had come. A year and a half of burning impatience for Manya. But it had taken her that long to save enough money to start her studies at the Sorbonne. Now, at last, the moment of departure had arrived. The train was waiting.

Manya had sent her mattress, sheets, blankets, and towels to Paris by freight. She had tried to think of everything she would need. There would be no money for extras once she left home.

A large wooden trunk, on which Manya had proudly painted the letters: M. S., contained underwear, three dresses, shoes, and two hats. The trunk had been sent on ahead, but Manya still had to carry a package of food for the three-day journey. She carried her scientific books also. And a quilt, because it was October and the railway carriage would be cold. And a small folding chair to use in the German fourth-class carriage, which would not be much more than a freight car. And a little bag of caramels that the professor had slipped into her coat pocket.

The train whistle gave a warning toot. Professor Sklodovski started nervously.

"Are you sure you have your train tickets, Manya?" he asked. "And your passport so that you can go through Germany and France with no trouble?" he added anxiously.

"Yes, Father," Manya felt in the side pocket of her big, shabby coat, "I have them right here. Third-class ticket to the German border, and fourth-class ticket across Germany. And here's my passport."

The whistle gave a long, shrill blast.

"Oh, Father!" Manya flung herself into his arms.

The professor kissed her tenderly. "Come back quickly," he whispered. "Work hard. Good luck!"

He watched her as she climbed the steps and entered the railway carriage. She turned and waved to him. Then he stood there until the train had pulled out of the station and disappeared from sight.

Meanwhile, on the train, Manya had stowed her

various bundles in the baggage net that hung over-head. She settled herself on the hard bench and tried to convince herself that she was really on her way to Paris at last. What would it be like, she wondered. What lay ahead?

She thought of her father returning to an empty house, and tears came to her eyes. "But it won't be for long," she told herself. "I will pass the examinations, and soon I will be back in Warsaw, teaching. And then I can take care of Father all the rest of his life."

She leaned her head back against the hard wall of the railway carriage and closed her eyes. Above the chatter of her fellow passengers, the iron wheels grinding along the tracks seemed to be saying: "PA-ris, Paris, PA-ris, Paris!" All of a sudden, Manya felt supremely happy.

Three days later, the train pulled into the Gare du Nord, one of the railroad stations in Paris. Manya and the other passengers rose wearily and stretched their tired backs. Everyone collected various belongings and climbed down from the railway carriage.

But once out of the station, Manya's spirits rose.

She was here! She was in Paris, where no Russian spies lurked, waiting to overhear an unwise remark. Where one could say whatever one pleased, in any language one chose to speak. Best of all, where one could learn!

Manya wished she could go straight to the Sorbonne. She did not think she could wait another moment to see the famous old university buildings. But she was still carrying the quilt and the books and the folding chair. So, in timid French, she asked a passer-by which omnibus would take her to 92 Rue d'Allemagne. This street was in one of the poorer sections of Paris, where Bronya and her husband lived.

When the double-deck omnibus came, drawn by three black horses, Manya climbed on eagerly. She tried to see everything as the horses plodded slowly along the street between rows of leafless elm trees. The shops, the people, a glimpse of cathedral towers, the beautiful river Seine flowing through the city—everything seemed wonderful to her.

When Manya reached Bronya's home, her sister's husband, Dr. Casimir Dluski, was there to greet her. Bronya herself was away but was expected back shortly. Casimir was a handsome and charming young Polish gentleman. He, like Bronya, had become a doctor. And now they shared the little consulting room in their apartment. Both young doctors worked long hours for little money. But they were very happy together.

"Welcome, little sister," Casimir said now, as he ushered Manya into the attractive living room. "As you know, your father has made me responsible for you. I shall expect you to show me great respect." The gleam of mischief in his black eyes betrayed him. Manya knew they would get along perfectly.

"You and Bronya are very good to allow me to come to you, Casimir," she said earnestly. "I promise I shall not bother you."

"On the contrary. You will be an ornament to our home," he replied gallantly. "Come now, let me take your things, and I will show you the room Bronya has prepared for you. It is small," he added as they walked down the hall. "But Bronya thought it would be the quietest place in the house, and you would be able to work better."

He opened a door. "Here we are," he said. "I shall have to leave you now, as my office hours are from one to three. And then I have some calls to make. But we will dine together."

"Casimir," Manya said, "before you go, will you tell me, please, how to get to the Sorbonne? I would like to go there and register for the courses."

Casimir gave Manya careful directions. And about an hour later, she was stepping off the omnibus and approaching the iron gate of the university. A large white sign on the wall near the entrance said:

FRENCH REPUBLIC
Faculty of Sciences—First Quarter
Courses will begin at the Sorbonne on
November 3, 1891

Manya sighed with joy and entered the building. When she came out, she was no longer Manya, or Marya. When she registered, she had spelled her name the French way, *Marie*—Mademoiselle Marie Sklodovska, student in the College of Science.

CHAPTER SEVEN

"First in Order of Merit"

A COLD March wind rattled the window of Marie's little room at the end of the hall, but she did not hear it. Studying at her desk under the lamplight, Marie held her thumbs over her ears as she used to do as a child.

But she could not help hearing the burst of laughter that echoed down the hall from the living room. Almost every evening, friends dropped in to see Bronya and Casimir. Casimir would play the piano, and Bronya would serve tea and little sweet cakes.

With a sigh, Marie straightened in her chair and stretched.

"Oh dear," she thought. "They're both such darlings, it's no wonder they have so many friends. But how am I ever going to keep my mind on my work with all this noise every evening?"

She sat forward again and turned a page in her book of physics. But even as she did so, she heard a man's quick footsteps coming down the hall.

Marie groaned, as a knock came at the door. "Come in," she called, a bit impatiently.

The door opened, and as she had expected, there stood Casimir.

"Now, Casimir," Marie began severely, "you know I have to—"

"Not a word, little sister!" Casimir interrupted,

his black eyes dancing. "I knew I would be scolded for disturbing you, but this you must not miss. Young Ignace Paderewski is here, and he is going to play the piano for us. You know how marvelously he plays. You must come."

"He does play beautifully," Marie agreed. "Well, I would hear the piano anyway, so I might as well go in with you and enjoy it."

She closed her book and rose from the desk. There would be no more chance to study tonight.

But the next evening, as Marie sat with her sister and brother-in-law at dinner, she said, "I have been thinking, perhaps I ought to move to the Latin Quarter. Many poor students live in that section of Paris. It is within walking distance of the library and the laboratory. I would save carfare and two hours of traveling time a day."

"But Manya," said Bronya, who could not get used to calling her sister Marie, "you would have to pay board and lodging if you left us. How could you possibly manage it?"

"I have walked around the neighborhood over there," Marie told her. "One can rent an attic room for very little. And if I stretch my savings, I will have about forty rubles a month to spend."

"Forty rubles!" Casimir roared with laughter. "Little sister, that is practically *nothing* a day."

"Nevertheless, I think I must try it," Marie said stubbornly.

Bronya and Casimir knew that if Marie had made up her mind, she could not be swayed. But they insisted on paying for the moving.

Bronya hired a boy with a handcart to take Marie's things to the room she had found. A folding bed, a mattress, a wooden table and chair were loaded into the cart. A washbasin, an oil lamp, and an alcohol heater were stowed in also. Space was found for two plates, three glasses, a knife, fork, spoon, and cup. And of course the teakettle.

When everything was ready, the boy set out. Then Bronya and Casimir took Manya by omnibus to her new home, which was in the attic of No. 3 Rue Flatters, not far from the university.

"Please, please, take care of yourself," Bronya begged as she kissed her sister good-by.

"And let me know if anyone gives you any trouble," Casimir said sternly. He still did not approve of Marie living by herself.

"I will," Marie promised, hugging them both. "And thank you, thank you for everything."

"We'll come and see you some evening soon, after you've had a chance to get settled," Bronya called as they started down the street.

Now Marie was truly alone. And she settled down to a program of study that scarcely gave her time to eat and sleep.

To save coal and oil, she studied each evening at the Library of Sainte Geneviève until it was closed at ten o'clock. Then she would come home and study in her bare, unheated room until two in the morning. It was only when her fingers grew too cold to hold her pen, and her eyes too tired to read another page, that she finally went to bed.

She was out in all kinds of weather. She lived on bread and butter and tea, with once in a while a bit of chocolate or a piece of fruit—if she happened to think of it. It is no wonder that the six flights of stairs began to seem like a mountain to climb.

Bronya and Casimir knew nothing of all this. When they noticed that Marie looked very tired, she would say she was overworked. But one day, returning from the Sorbonne, Marie climbed five of the six flights of stairs, and then she fainted.

A classmate, who lived on the fifth floor of the house, found her on the floor in the hall. She helped Marie upstairs to her room and then hastened to Bronya's house to tell her Marie was ill.

Within a short time, Casimir arrived. He noticed the unused plates and the lack of groceries on the shelf. Helping Marie put on her hat and coat, he took her back to the apartment in the Rue d'Allemagne.

"Bronya," he called as he unlocked the front door. "The little one is here, and what she needs is a large, rare steak—immediately."

Something in her husband's voice brought Bronya running. She took one look at Marie's white, thin face and without a word went to the kitchen. In twenty minutes, Marie was sitting at the dining room table slowly eating beefsteak and crisp fried potatoes.

As color began to come back to her face, Casimir eyed her sternly.

"Now, my little sister," he said, "you will be good

enough to tell us what you had for luncheon today."

"Why, I—" Marie stopped. She really couldn't remember whether she had eaten or not. "Oh yes," she said. "I—well, I had some radishes."

"So! You had some radishes!" Casimir said sarcastically. "And what, may I ask, did you have for dinner last night?"

"Dinner? Oh. Well, that was radishes, too, and some cherries."

"Oh, Manya!" Bronya exclaimed, horrified.

"Manya, I am very angry with you," Casimir said, "but I am even more angry with myself. I promised your father I would look after you, and I have failed to do so. However, we will do things differently now. You are to stay in bed here for a week, and eat as often and as much as I say you must."

"But Casimir," Marie protested in dismay. "The examinations are approaching. I must not miss the classes."

"You may study here at home," Casimir said, "but you are not going to get up until you have regained your strength."

"Casimir is right, dear," Bronya said gently.

"You are so weak now that you couldn't pass an examination."

Marie realized that they were right. Meekly, she let them feed her and make her go to sleep early each night. At the end of a week she felt much better. When she promised to take better care of herself in the future, Bronya and Casimir allowed her to go back to her attic.

And for a while Marie did try to live more sensibly. But she was making exciting progress in her studies. Cold, hunger, and weariness honestly did not seem important to her. Professor Lippmann, in charge of the physics laboratory, noticed that his quiet little Polish student had a very fine mind. He gave her certain work to do which allowed her to try some ideas of her own.

In the high-ceilinged, wide physics lab, Marie in her linen smock would stand in front of a wooden table. She worked with fine instruments and test tubes. Alongside of her other young students worked. No one spoke, for each student was intent upon what he was doing. As long as she lived, Marie was to love best the silence of the laboratory.

When April came to Paris, even Marie could not

stay indoors. The horse-chestnut trees sprang into full bloom, and the scent of lilacs was everywhere. On Sundays, she climbed to the upper deck of the omnibus and rode out into the country. In the woods she picked wild flowers and brought them back to brighten her miserable little room.

But as July of 1893 approached, Marie took no more excursions to the country. The examinations for the Master's Degree in Physics were coming. She spent every moment studying.

Finally the awful day arrived. Marie and her thirty fellow-students filed into the hall for the first test. The weather by this time was dreadfully hot.

Marie slipped into the seat behind the desk assigned to her, and picked up the examination papers. She read the first question, and her mind was a blank. The heat, combined with her own nervousness, had driven every thought out of her head.

She read the second question, and it, too, meant absolutely nothing.

"Whatever am I to do?" Marie thought in panic. "I don't know *any* of the answers."

For a moment she sat there trembling. Then she

said to herself sternly, "You are not going to let yourself be beaten down by this or anything else. Now, get to work!"

Her panic subsided, and Marie got to work . . .

Ten days later, Marie sat in the crowded amphitheater waiting to hear the results of the examinations. Other students sat all around in the great circle of seats, chattering nervously. A sudden, solemn silence fell as the tall, gray-bearded examiner entered the arena.

In the thrilling hush, his deep voice rang out grave and clear:

"First in order of merit, Mademoiselle Marie Sklodovska."

CHAPTER EIGHT

Baby Irene

MARIE walked briskly along the dark, wintery street. She was thinking of the happy summer months she had spent at home with her father and other relatives. And she was thinking, too, of a scholarship she had won.

"Six hundred rubles!" she thought happily. This was about three hundred dollars. "Without that money I could never have come back to the Sorbonne for a second year. The mathematics course I am taking seems terribly hard. But then, so did physics, at first."

She stopped at a crossing to let a horse-drawn cart pass noisily on the cobblestoned street. Then, turning up her coat collar, she shivered a little and hurried on her way to visit some Polish friends.

When she reached the boardinghouse in which her friends were staying, she found another guest there before her. He was a young scientist named Pierre Curie.

Pierre Curie was a tall man, a few years older than Marie. He had fine hands and honest, peaceful eyes. His slow, friendly smile made Marie like him at once.

The four young people sat down near the balcony window, through which the lights of Paris twinkled. They drank tea and talked science. Pierre Curie was amazed at how much this beautiful young Polish student knew.

"Are you going to make France your home?" he asked her.

"Oh, no," Marie replied. "I shall take my mathematics examination this summer. Then, if I can manage it, I'll come back for one more year. After that, I shall become a teacher in Poland. We need teachers there."

"But you would not be able to continue your scientific studies in Poland," Pierre objected. Even as he said the words, he knew what he really meant: he did not want Marie to leave France—ever.

In the spring months that followed their meeting, Pierre gently persuaded Marie to see him more and more often. By July, she had passed her mathematics examination and was about to go home to Warsaw. Pierre asked her to stay in Paris and marry him. But Marie could not bear the thought of leaving her family and Poland forever. She promised only that she would try to return to Paris in the fall.

Marie kept that promise. This time she came back to live in a small room next to the office Bronya had taken in the Rue de Chateaudun. It was here that Pierre met Bronya and Casimir, who liked him very much. And it was here that Pierre made the suggestion that probably won Marie's heart.

Bringing her home from a day in the country, Pierre said:

"If I were to leave Paris, go to Poland and get a position then, *then* would you marry me? I could give French lessons, and we could do scientific work whenever we had the time."

"Pierre, how can you think of such a thing?" Marie exclaimed. "You are a genius. You must never give up your work here in Paris."

Marie meant what she said. And in the same moment, she realized she could no longer bear the thought of being away from Pierre.

On July 26, 1895, the sun rose in a cloudless sky. Marie brushed her ash-blond hair carefully, then pinned it high on her head. She put on a dark blue suit with a lighter blue, striped blouse. The suit and blouse were a gift from Casimir's mother. For today was Marie's wedding day.

Pierre came for her, and they took a train to the town of Sceaux, where his parents lived. Professor Sklodovski was waiting at the Curies' house. Bronya, Casimir, and Hela were there, too, and a few friends who had been invited to the wedding.

After the simple ceremony, Marie changed into a white blouse and shorter skirt. Around her slender waist she buckled a leather belt with pockets, in which she placed her watch and a little pocket knife. She changed her dainty wedding slippers for stout walking shoes. Then she ran downstairs to join Pierre, who had also changed to country clothes.

Together they went out to the garden, where another wedding present waited—two brand-new shiny bicycles. Extra clothing and two long raincoats had already been packed in knapsacks on the bicycles.

Everybody came to the garden gate to kiss the young couple good-by and wave as the happy pair pedaled gaily off.

"There isn't a soul on earth to equal my Pierre," his mother said, as she watched their departing backs. "Marie will be happier with him than with anybody in the world."

"And you will have a loving daughter in Marie, madame," Professor Sklodovski assured the old lady.

Meanwhile, Marie and Pierre were bicycling along a country lane, feeling free as two birds.

They had packed a snack of bread and cheese, and when they began to feel hungry they stopped beside a quiet pond and ate their little lunch. Then Marie lay down under a giant tree with her arms under her head. She gazed at the blue sky until she was almost asleep.

Suddenly something cold and dripping touched her. She sat up with a little gasp, and saw that Pierre had dropped a little green frog into her hand.

"Oh-oh, Pierre, take it away!" she cried, as the little green frog looked up at her.

"You don't like frogs?" Pierre asked, amazed. But he took the little frog and put it back into the pool. "Then here is something you will like."

While Marie had been dozing, Pierre had gathered yellow irises and pale pink pond lilies for her.

"Pierre, they are beautiful," Marie said delightedly, and she put some of the irises in her hair.

Soon they got on their bicycles again and continued along the tree-shaded lane. When it grew dusk, they stopped at a little inn for the night. They had supper by candlelight—great steaming bowls of thick, hot vegetable soup with crusty hunks of delicious French bread.

The free, lazy days went by all too quickly. At last it was time to return to Paris and to work. Marie had decided to study for a diploma that would allow her to teach school in France. Pierre's position as chief of the laboratory at the School of Physics did not pay much, and she wanted to earn some money to help him.

A little apartment was found on the fourth floor of No. 24 Rue de la Glacière, with windows overlooking a garden. Marie and Pierre furnished it only with what was absolutely necessary.

The walls were bare except for the shelves of books. In the middle of the tiny living room they placed a long white wooden table. Marie sat at one end of it and Pierre at the other. In the center of the table were only the oil lamp and a bunch of flowers.

In the evenings, as Pierre made up the program for his winter classes, Marie studied for her teacher's certificate. She also wrote in a large, black important-looking book. The pages were headed: *Monsieur's Expenses* and *Madame's Expenses*. She kept track of every franc she spent.

And Marie learned to cook. She found this much harder than the most difficult of chemistry experiments. The beans would burn, and the chicken would stick to the pan!

Sometimes Marie thought longingly of the old days when all she did was butter some bread and brew a cup of tea. But she managed to do the marketing and housekeeping, to work eight hours a day in Pierre's laboratory, and to study five hours at night.

In September of 1897, baby Irene was born. Marie and Pierre were sure she was the most beautiful baby the world had ever seen. Marie bathed the baby four times a day. And Pierre came home with tiny baby jackets and shirts in two sizes. He was convinced that little Irene was growing before their very eyes.

Marie was forced to hire a nurse to take care of the baby during the day. She had won her teacher's diploma, but she was still very busy working in the laboratory every day.

One morning, having returned from the market with Irene and the vegetables in the carriage, Marie prepared to leave for the lab. But first she weighed the baby and carefully noted her weight in a little gray notebook. Then, when the nurse arrived, she hugged Irene and placed her back in her carriage.

"It is a fine, sunny day," Marie told the nurse. "You will take her to the park early?"

"*Oui*, madame," the kindly woman said, smiling.

Marie put on her plain little black hat and her shabby coat, and left the apartment. She walked briskly the short distance to the School of Physics. Soon she was deep in a difficult experiment. But about eleven o'clock, a feeling of panic swept over her. What if someone had stolen the baby when the nurse was not watching!

"No, do not be ridiculous," Marie said to herself. But what if someone *had!*

Without stopping to put on her hat, though the November day was cold, Marie rushed out of the laboratory. She hastened up the street toward the Parc Monsouris. Her eyes searched anxiously for the familiar carriage.

Then she sighed with relief. For there was the baby in her warm little white coat, with the nurse sitting faithfully beside the carriage.

Feeling very, very foolish, Marie returned to the laboratory.

CHAPTER NINE

An Exciting Discovery

IT WAS a windy night in January. Supper was over, the dishes and pans washed, and four-months-old Irene had been put to bed. Marie tiptoed out of the baby's room and slipped quietly into her chair at one end of the living-room table. She did not wish to disturb Pierre. He was deep in his study of crystals on which he had been working for some time.

Marie picked up a magazine article which described a new scientific experiment and reread it carefully.

About two years earlier, a German scientist, Wilhelm von Roentgen, as an experiment, had sent an alternating current of electricity through a glass tube. The tube was nearly a vacuum, that is, it had almost no air inside it. The tube had been covered with black paper, so the light within it was completely hidden. But Roentgen noticed that some crystals on a table near by were shining. He did not know what to make of this, since ordinary light rays do not go through thick black paper. Yet the crystals were shining. What was coming from the tube to make them shine? He didn't know. So he called the mysterious rays X-rays.

A French scientist, Professor Henri Becquerel, had read about Roentgen's experiment. Becquerel decided to find out if certain substances which became radiant in sunlight were giving off these same

X-rays. He experimented with a rare metal called uranium.

Then, quite by accident, Becquerel discovered something entirely different from what he had expected to find. He had left a uranium sample next to a wrapped photographic plate in a dark drawer of his desk for several months. When he looked at the plate later, he found a photographic image on it.

Uranium, then, gave off its strange rays even when it had *not* been exposed to sunlight or to the alternating electric current that produced X-rays!

Marie and Pierre Curie had read about these experiments with great interest—Marie particularly, for she was determined to learn as much as she could about physics.

Tonight, she waited until Pierre raised his eyes from his book and glanced over at her. Then she said eagerly, "Pierre, I have an idea. I should like to find out what makes uranium and other substances with uranium in them give off those strange rays."

"I think that is a splendid idea, Marie," Pierre exclaimed. "I don't think anyone so far has followed up Professor Becquerel's discovery. And you might find out something important that would add to our knowledge of science."

Marie frowned thoughtfully. "I shall have to find some place to work," she said. "I'll have to test a great many samples, and I'll need more space than I have in the school lab."

Pierre nodded. "Let me speak to the school director," he suggested. "Surely there must be some room in the school that he would be willing to let you use."

The next morning Pierre did ask the director of the school. But the only place the director could offer Marie was a small, glass-walled storeroom on the courtyard of the building. The room was damp and drafty and full of old lumber and rusty machinery. But since Marie had no choice, she took it.

Three days later, she was ready to go work. She had very little equipment to work with. The instruments she did have sometimes did not register correctly because of the cold wind that came through the cracks under the windows. One day in February the temperature dropped to barely six degrees above zero. But Marie put on a sweater under her smock and kept on working.

AN EXCITING DISCOVERY

By the end of March, she had tested enough chemical samples to be sure of three things. The more uranium there was in the sample, the more powerful were the mysterious rays. Light, heat, and cold had nothing to do with the power of the rays. Nor did it matter if the uranium was combined with something else.

"Pierre," Marie said, as they walked home from the school together, one rainy evening. "These rays that come from uranium—they are not caused by light that we can see. Nor by heat that we can feel. I think they must come from light, or heat, that was imprisoned within the uranium when the earth was new and very hot."

Pierre thought about this a moment.

"You may be right," he agreed.

"If I am right," Marie went on eagerly, "perhaps this same radiation happens in other metals, not just uranium."

Pierre smiled down at her. "It would take a lot of work to find that out," he said. "You would have to examine every single chemical element that we know anything about."

"That's just what I'm planning to do," Marie said.

Pierre whistled. "You are a true scientist!" he exclaimed.

But that was exactly what Marie did. And the next few weeks' work proved that her guess had been right. She discovered that thorium, another substance that glowed, gave off rays just as uranium did.

"Pierre," Marie called excitedly, as she ran to meet him in the courtyard at the end of a day's work. "I've thought of a name for this strange radiance of uranium and thorium. Radioactivity! How do you like it?"

"Radioactivity," Pierre Curie repeated thoughtfully. "And substances that give off these rays are *radioactive*. Yes, I like that name."

"Do you know what I'm going to do now?" Marie was as happy as a child opening a birthday present. "I'm going to examine all the minerals I can get hold of, and see which ones are radioactive. And then, with the electrometer, I'm going to measure the power of the rays in each of those samples. And I'm going to find out how much uranium or thorium there is in each sample."

But when Marie did this, she ran into a mystery.

"I *must* have made a mistake somewhere," she thought one day, as she stood in front of her work-table in the little glass-walled storeroom. "I know exactly how much uranium there is in this sample of pitchblende ore. It couldn't possibly cause as much radioactivity as the electrometer shows."

She patiently repeated the experiment, not once but many times. Still the results showed much more powerful rays than could have been caused by the amount of uranium in the sample of ore.

Marie was very tired when she came home that night. But she could not get the puzzle out of her mind. And when Bronya stopped in after dinner, Marie told her sister about it.

"I simply can't understand it," Marie said. "I've been working with pitchblende, which, as you know, is a uranium ore. In Austria, where they mine it, I believe they take out the uranium to use in making glass. Well, I know now just how much radio-activity is caused by a certain amount of uranium. But every time I do the experiment, there's much more radioactivity than there should be."

"And you can't explain why that is?" Bronya asked thoughtfully.

"No. I'm beginning to think—" Marie turned to smile at Pierre, who sat nearby. "Don't laugh, Pierre, but I'm really beginning to think some *unknown* chemical element must be causing these more powerful rays."

Pierre leaned forward, his eyes lighting with interest. "I am not laughing, Marie," he said. "After all, a scientist is like an explorer. When Columbus set out for India, he found, instead, the great unknown continent of America."

"Marie!" Bronya exclaimed. "Just think! If you're right and you should find some new, unknown element—why, it would be one of the most important discoveries ever made in science!"

CHAPTER TEN

Radium!

SPRING had come to Paris again. Marie opened the bedroom window one Sunday morning, and breathed deep of the soft air. She looked out over the city where tall, misty green elms traced the streets. In the distance she could see the lofty Eiffel Tower, and everywhere the river Seine with its green banks.

She went over to Irene's crib and took the baby up in her arms.

"You're such a big girl now," Marie cooed, kissing her. "You're eight months old today. Imagine that!"

"Oh, there you are, darling," Pierre said, coming into the bedroom. "I was wondering if you'd like to take a walk by the river before the day grows too warm."

"Yes, let's," Marie agreed. "I will give Irene her bath when we get back. And I'll have time to pare the vegetables for dinner before I go to the lab."

"Today," Pierre said, smiling, "I am going to the lab with you."

"You are?" Marie exclaimed in pleased surprise. "But on Sunday you always work on the crystals in the—"

Pierre laughed. "I don't believe you quite realize the importance of your work even now, Marie," he said. "But other scientists do. One of them has even written an article about the unknown element you think you may have discovered in pitchblende ore. So I've decided to stop my experiments with crystals for a while, and help you try to find it."

"Pierre, that is wonderful!" Marie said, as she started to dress the baby.

Soon they were ready, and the tall, bearded young man and the slender, blond young woman strolled down the Rue de la Glacière. Marie was pushing the baby in the carriage, and they looked like any other young couple taking an early morning walk. But what they talked about was quite different. For Marie and Pierre already were planning how they would work together in Marie's little laboratory.

From then on, it was "We found," or "We observed," in all their notebooks and reports. By July, Marie and Pierre were sure that there was not just one unknown element in pitchblende ore. There were two.

When they finally succeeded in chemically separating one of the new elements, Pierre said to Marie, "You must name it."

Marie thought a moment. Then she suggested, "Could we call it—polonium?"

Pierre smiled. He knew that in selecting that name, Marie was thinking of her beloved Poland.

Paris had grown unbearably hot by this time. Marie and Pierre rented a little peasant's cottage in the country. Then they closed the apartment in Paris and took a vacation.

In the country, Irene grew rapidly. She was beginning to try to stand up all by herself. She wanted to chase Didi, the alley cat that had adopted the Curie family.

"Gogli, gogli, go!" Irene would scream. But Didi always scampered neatly out of reach of the chubby little hands.

The happy summer days of swimming, climbing, and bicycling were marred by only one thing. Bronya and Casimir had decided to leave Paris and build a hospital in the Carpathian Mountains.

When the time came to say good-by, Marie and Bronya both felt very sad. They had always been close friends as well as sisters. They knew they would miss each other very much.

Marie was almost glad when September came and she and Pierre could return to Paris, and to work. By December, the two young scientists were able to announce that they had located the second unknown element for which they had been searching.

They called it radium.

But it was one thing to say that radium existed. It was quite another thing to convince the scientists who were watching Marie and Pierre's work with keen interest. If there really were such a thing as radium, these men argued, then they wanted to see it, and to weigh it.

"We shall have to produce salts of actual radium," Marie said to Pierre.

Pierre nodded. Then he added, "But that means we'll need a lot of pitchblende ore, tons of it. We can never afford to buy it— Wait a minute!" he exclaimed. "You mentioned once that at the mines in Austria they take out the uranium to use in making glass. What is left of the pitchblende is of no use to the miners. Perhaps we can get hold of some of it cheaply. Let's try."

Once Marie and Pierre had decided on a plan, they wasted no time in carrying it out. And one morning a big wagon loaded with burlap sacks filled with pitchblende pulled up in the courtyard of the School of Physics. Marie and Pierre rushed out to the wagon.

"Oh, do cut the string and open one of the sacks," Marie begged Pierre, "so we can look at our pitchblende."

Laughing at her excitement, Pierre opened one of the sacks. Delightedly Marie thrust her hand in and pulled out a fistful of the dull, brownish ore. In the meantime, Pierre told the driver to unload the sacks and take them into a shed on the opposite side of the courtyard.

This old shed was the only place they had been able to find that was big enough to work in. It had an earth floor and flaking plaster walls. When it rained, the roof leaked. The only heat was provided by a rusty old iron stove. But it did have a blackboard, some old tables, and a dusty skylight in the roof.

Marie and Pierre had moved their equipment from the little workroom to the shed. Now they were eager to start work. They might not have felt so eager if they had known it was going to take fortyfive months of back-breaking labor to find what they sought.

In the meantime, several things happened. Pierre's mother had died, and his father, old Dr. Curie, came to live with Marie and Pierre. The old gentleman was tall and had lively blue eyes that snapped when he got into an argument—which was quite often. But he was the soul of kindness and he adored little Irene.

When the little girl was about three years old, the Curies moved to a small one-story house on the Boulevard Kellermann. It was screened from the street by thick, tall trees. Behind the house was a quiet garden. Here Irene could play safely, under the watchful eye of her grandfather.

It meant a great deal to Marie to know that old Dr. Curie was never far away from Irene. For, from nine in the morning until the dark came, Marie was busy.

Day after day she stood in the windy courtyard of the school, stirring with a long iron rod the boiling chemicals which she hoped someday would yield salts of radium. She mixed chemicals, and she separated them, filling great jars with strange-looking substances which she and Pierre would study.

After nearly three years, even Pierre, who worked inside the shed, became discouraged.

"Marie, Marie," he exclaimed one day, when she had stopped work long enough to come indoors and drink a cup of tea with him. "You cannot go on like this. It will take years to try to prepare radium in this old shed, which is always either too hot or too cold. You may never get it. Why not give up this operation and simply continue your study of radio-activity?"

"No," Marie said stubbornly, "I will keep on until I get our radium. I wonder what it will be like," she added dreamily.

"Perhaps it will have a beautiful color," Pierre said, smiling.

"I must get it, I must . . ." Marie murmured.

On a May evening in 1902, Marie and Pierre walked home from the shed in the gathering twilight. They had worked hard all day, but their heads were high, and their steps were light. Little Irene, four years old now, came running to meet them. "Mé, Mé," she called, her baby way of saying, "Mother," and flung herself into her mother's arms.

Marie hugged the child, and taking her hand,

went into the house. Soon she had supper on the table, and the little family sat down to eat. Old Dr. Curie had tied a bib on Irene, and the little girl was busy picking the "pearls" out of her tapioca pudding.

Marie looked across the table at Pierre. Her gray eyes were sparkling. Pierre turned to his father, who sat opposite Irene.

"Father," Pierre said, "Marie has something to tell you. Something we want you to know before anyone else knows."

"Yes?" The old doctor looked at Marie curiously. "What is it, my dear?"

"We have done it!" Marie announced triumphantly. "We have finally obtained—salts of radium!"

"Marie! My congratulations to you both!" old Dr. Curie exclaimed. He rose from his chair and came and kissed Marie on both cheeks. Then he embraced his son also.

Soon after supper, Marie put Irene to bed. This meant first a good deal of splashing in the tub, several requests of "Mé, don't go away so soon," and finally, "Mé . . . may I have a glass of water?"

At last Irene was asleep. Pierre's father went to bed early also. When Marie came back to the living room, she found Pierre prowling around restlessly. Marie sat down in the old mahogany armchair and began to sew on an apron she was making for Irene. But she could not keep her mind on her sewing, and pricked her finger twice. Finally she put down her work and looked at Pierre.

"Shall we go back—" she began.

"Yes, let's," he agreed. She did not need to finish her sentence. Both of them were longing to return to the lab.

In the dingy school courtyard once again, Pierre pushed open the squeaky door of the old shed. They walked into the dark workroom.

But now the old shed had become beautiful. For in the darkness, like the luminous hands of a clock in the night—but far brighter—glowed and shimmered the radiant, bluish fragments of precious radium!

CHAPTER ELEVEN

An Important Decision

I<small>T WAS</small> the morning after Marie and Pierre had gone back to the shed to look at their radium. Marie sat down happily at the table to write her father the news. She knew he would be delighted, for the old professor had followed each step of her work with great interest.

In a few days, a reply came to her letter. Her father was overjoyed. Unfortunately, he was too ill to make the journey from Warsaw to Paris to see the extraordinary new element. But Marie's news had made him feel much better.

Then the blow fell. Marie received a telegram from her brother Joseph bidding her come to Warsaw at once. Her father was dying.

Marie left on the first train she could get. But the journey took nearly three days. When she reached Joseph's house, her father was gone.

Marie sobbed bitterly in Bronya's arms. "I should never have stayed in Paris," she cried. "He always wanted me to come home and live with him."

"Manya," Bronya said soothingly, stroking her sister's fair hair. "Do not feel that way. Father was so proud of you. Your discovery of radium gave him great happiness."

This comforted Marie a little. But it was a pale, tired young woman who returned to Paris. The shock and grief of her father's death, coming on top of four long years of work, left her nearly worn out.

Added to that, she had begun teaching in a girls' school in Sèvres. She had to travel there by train several times a week. But the worst worry of all was Pierre.

In order to make more money, Pierre had taken a position teaching physics at a branch of the Sorbonne. He, too, was tired out and often suffered pain. Marie was terrified lest this pain be caused by some serious trouble.

"Pierre," she said one day, when he seemed particularly tired. "We—we can't exist without each other, can we?"

Pierre started to say, "No." Then he looked at her. "You are wrong," he said. "Whatever happens, even if one has to go on like a body without a soul, one must work just the same."

It was becoming very hard now for Marie and Pierre to work without interruption. The announcement of their discovery of radium had caused a great deal of excitement among scientists all over the world. Up to that time, it had been believed that the world was created of certain definite substances. These substances, or elements, were supposed to be made up of certain atoms, the nature of which never changed.

Now, Marie and Pierre had proved that some of these atoms were changing all the time. With tremendous energy, tiny pieces kept breaking off from the heart, or nucleus, of each atom.

This breaking off was what caused the strange rays that Henri Becquerel had first noticed in uranium. And when enough particles had broken off from an atom of uranium, it was no longer uranium. After passing through several stages, it had become an atom of radium.

This breaking-off process never stopped. An atom of radium would in time become an atom of polonium. This also would change, and finally would turn into lead.

There was another reason why the Curies had many visitors who interrupted their work.

It had been discovered that radium was useful in many ways. It could be used to tell real diamonds from imitations. It gave off heat that might someday be used in place of coal or oil. And it had another use which was the most important of all.

One afternoon, a visitor came to the shed where

Marie and Pierre were working. It was Professor Henri Becquerel, who had become a good friend.

"What do you think your radium has done to me?" he demanded. "Look at my waistcoat pocket! I had a tube of radium in there, and it's burned a hole right through my shirt. I might add, it burned *me*, too!"

Marie and Pierre laughed. "We get burned all the time," Pierre said. "After about two weeks, you'll have a blister, and it'll take about two weeks more to clear up."

"But aren't you afraid of what this radiation may be doing to you?" Professor Becquerel asked, astonished. "Working with radium as closely as you two do may ruin your health."

Marie shrugged. "That's a chance we have to take," she said calmly.

"Yes," Pierre agreed, "we can't worry about that. But I would like to find out what effect these powerful rays might have on people with skin diseases and tumors."

"You mean, if radium can burn healthy skin, it could also burn away diseased skin." Becquerel nodded vigorously. "I think you should certainly look into it."

Not long after, Pierre and some other scientists treated sick animals with radium. They were happy to find that they could cure the animals. As a result of their experiments, they discovered that radium could be used to help certain forms of cancer.

Now the value of radium soared. One gram was worth 750,000 gold francs. At that time, this sum was equal to $150,000. All over Europe, and in England and America, many people were anxious to produce the rare metal that had become so valuable.

One morning a letter came, addressed to Pierre Curie. It was from some businessmen in the city of Buffalo, in the United States. Pierre read it carefully. Then, holding the letter in his hand, he went to find Marie, who was in the bedroom.

Marie was trying to make five-year-old Irene stand still while she brushed her light brown hair. But Irene was much more interested in trying to tickle Didi, the black-and-white cat.

"Marie, I'd like to talk to you about something,"

Pierre said. "Something important that has come up about radium."

Marie glanced at his thin, kind face and saw that he was troubled.

"Run along then, Irene," she told the bright-eyed little girl. "I wasn't getting anywhere with you anyway."

With a shout of triumph, Irene pounced upon the surprised Didi. She caught the cat up in her chubby little arms. The two rolled on the wide bed like a puppy and a kitten playing together.

Laughing, Marie followed Pierre into the living room.

"This letter," he told her, sitting down at his desk, "is from some people in America. They want us to tell their engineers how you managed to get salts of pure radium from pitchblende ore. They want to be able to do it, too."

"So?" Marie murmured. She was still trying to think what could be troubling Pierre.

"Well," he went on, "we can *give* them the information they want or—" He paused. "Or we can claim the right to *sell* the information to them. After all, it was you who found the way to get the radium out of pitchblende. And you and I are the only ones who know how you did it."

"But Pierre." Marie was troubled now, too. "Even though it is our discovery, doesn't it belong to the world? Radium is going to be used to cure painful diseases. Have we the right to make money on that?"

"Perhaps not," Pierre said soberly. "But at the same time, we must remember, we live a hard life, Marie. We have no money to build a proper laboratory. And we have little Irene's future to think of. If we were paid a certain sum of money every time we told someone how to manufacture radium from pitchblende—we would soon be rich."

Marie sat for a moment in silence. Then she rose from her chair and went to the window. "Pierre," she said finally. "Even if we shall always be poor, I don't see how we can take money for the information these people want. It would be contrary to the scientific spirit."

Pierre stood up and moved to her side. "You are right," he said, putting a hand on her shoulder. "It *would* be contrary to the scientific spirit. But I had to be sure you felt that way, too."

"Oh, I do, Pierre. I do," Marie assured him. She felt happy now that their decision had been made. "And you will write to the American engineers tonight."

"Yes. I will give them all the information they wish," Pierre promised. "Look!" he exclaimed gaily. "The trees all over the city have grown thick and green. Marie, let's—"

Marie laughed up at him. "Yes, let's!" she agreed.

Half an hour later, the bicycle tires had been pumped up. And Marie and Pierre Curie, like two children let out of school, headed for the woods.

CHAPTER TWELVE

A Penny for Irene

MARIE CURIE, in a simple, dark evening dress, glanced down the long banquet table. The English ladies seated between the gentlemen at the dinner wore such beautiful jewels that Marie could not take her eyes off them.

But, although Marie herself wore no jewelry, everybody else's eyes were on her. She and Pierre were the guests of honor at this banquet. They had come to England at the invitation of the Royal Institution. This important group of scientists had asked Pierre Curie to give them a lecture on radium.

The lecture had been a big success. And now everyone in London wished to meet Professor and Madame Curie and entertain them at parties and dinners.

Marie turned to Lord Kelvin, an old friend and fellow scientist, who was seated beside her. "Everyone is so kind," she said to the old gentleman. "Pierre and I are not used to having a fuss made of us. We rarely go out at all. We never seem to have the time."

"I can understand that," Lord Kelvin said, smiling. "But you must remember that you are now a very famous lady. You are the first woman who was ever invited to the sessions of the Royal Institution. In fact, I would say that right now you are the only famous woman scientist in the world."

Marie looked a little shocked when Lord Kelvin said that. She didn't feel famous—and she didn't want to be. It would take too much of her time. She glanced at Pierre.

He had just given a magnificent talk on radium. Now he was quietly answering the questions of the famous scientist who sat near him. Even in his rusty black evening clothes, it seemed to Marie that he was the most distinguished-looking man in the room.

In the carriage going back to their hotel, Pierre started to chuckle.

"What are you laughing at, *mon cher?*" Marie asked him.

"I saw you looking at the ladies' jewels," Pierre told her. "I, too, was looking at them. But not with so much admiration as you did. I was playing a little game.

"I was trying to guess how much money each necklace and bracelet was worth. And how big a laboratory I could build if I had that much money to spend. And do you know, Marie," he laughed, "before I knew it, I had a group of buildings as large as the Sorbonne!"

Marie laughed, too, but she patted his hand tenderly. "Pierre, you *must* have the laboratory you have wanted for so long," she sighed. "All these honors are very nice, but we would so much rather have a decent place to work in. Someday we'll get it, won't we?"

"Yes," agreed Pierre. "I'm sure we will."

In November, a few months after their return to Paris, Pierre made another trip to London. This time it was to receive the Davy Medal. This great honor had been awarded to Pierre and Marie by the Royal Society of London.

When he came home, Marie and little Irene met him at the door of the apartment. Irene, who had been told that her father was bringing home a prize, danced up and down with excitement.

"Papa!" she cried, after he had lifted her up and kissed her. "Let me see it, let me see it!"

Her father rummaged in the valise he had been carrying. "Now where did I put that thing?" he exclaimed. "I'm sure I brought it home with me—I think."

"Is this it, Pierre?" Marie had picked up a rather heavy package that lay under Pierre's hat on the table.

"Yes, that's it." Pierre unwrapped the package and lifted the leather lid of the case. He held up a heavy gold medal on which were engraved the names: PIERRE AND MARIE CURIE.

"Why, it's a great big gold penny!" Irene cried. "It's pretty, isn't it, Mé?"

"Yes, it's very pretty," Marie answered. "But what on earth will we do with it?"

"Perhaps we could put it on the table here, for a paper weight," Pierre suggested. "Or maybe we could—well—hang it on the wall."

He took up the medal to see how it would look over the mantel. It slipped out of his hands and fell to the floor with a thud.

Irene darted to pick it up and stood feeling of the ridges in the engraving. Then, laughing with glee, she began to roll it along the carpet like a hoop.

A happy smile spread over Pierre Curie's gentle face. "See, Irene loves her new toy," he said to Marie. "So, we have found a use for the medal after all."

Honors still continued to fall thick and fast upon the Curies. And a month later came the greatest honor of all. On December 10, 1903, the Academy

of Science in Stockholm, Sweden, announced that
the great Nobel Prize in Physics had been awarded
to Henri Becquerel, and to Pierre and Marie Curie.

Marie was delighted. For the Nobel Prize was
a cash prize of 140,000 francs. Of that money,
70,000 francs would be hers and Pierre's.

"Now, Pierre," Marie said, on the cold January
morning when the check came in the mail from
Sweden, "you can give up teaching at the School of
Physics, and you will not become so terribly tired
any more."

"And we can hire a laboratory assistant, Marie,"
Pierre added hopefully. "And you must give up your
position at the Sèvres School."

"No," Marie said, shaking her head stubbornly.
"I can manage the teaching. But Pierre," she asked
timidly, "would it be all right if I sent Bronya some
money? Just as a loan, but they do need help right
now with the hospital they are building."

"Of course," Pierre said. "And would you mind
if I helped out my brother with a little money, too?"

"Naturally, Pierre. And perhaps I might have
Irene's room repapered. It does need it."

Pierre smiled at his blond, slender wife.

"Wouldn't you like a new evening gown?" he asked. "Evening clothes become you so well."

"Oh, no," Marie protested. "My black silk is perfectly good. We mustn't spend the money foolishly."

It was wonderful to have some money to spend. But at the end of 1904, something happened that meant more to Marie and Pierre than all the honors and prizes they had won. On December 6th, a beautiful baby sister for Irene was born. Little Eve had great blue eyes and thick dark hair. Irene, who was seven now, came to the side of the crib to stare down at the baby.

"Is she not beautiful, Irene, your baby sister?" her mother murmured.

"Yes, Mé. She is like a little doll. Mé," Irene added, "will you bring me home some bananas?"

Marie laughed and ruffled Irene's hair. "I will bring you some soon," she said. "But now you must help me take care of little Eve."

Another thing brought happiness to the Curies that year. This was the appointment of Pierre as Professor of Physics at the Sorbonne University. He was to have a new laboratory and three assistants.

Best of all, Marie was to be his chief of laboratory at a salary of 2,400 francs a year.

The new laboratory, in the Rue Cuvier, was better than the shed they had been using. As soon as they could, Pierre and Marie moved all their equipment into the new lab and put everything in order. Then Pierre said, "Let's go back once more."

They walked back to the muddy courtyard. Slowly Pierre pushed open the squeaking door of the old shed, and they went inside.

"We put in long hours of hard work here, didn't we?" Pierre said, looking around at the rusty iron stove, the dirt floor, and the dust-covered glass roof.

"We did," Marie agreed. "But it was useful work, Pierre. And we've been very happy here."

After a few moments, they came out of the old shed together, and Pierre shut the squeaking door behind them for the last time.

CHAPTER THIRTEEN

"I Will Try"

AT St. Rémy, in the valley of Chevreuse, not far from Paris, Marie and Pierre had rented a quiet country cottage. Here, with Irene and little Eve—

and for the first time a servant—they spent the Easter holidays of 1906.

Sunday morning dawned fresh and clear. Everyone was up early. They had been awakened by Eve, now fourteen months old, who refused to lie in bed in the morning. After a good breakfast of sausage, milk, and fresh-baked country bread, Pierre made a suggestion.

"Marie," he said, "it's a perfect April day. Let's go for a bicycle ride. Irene and Eve will be safe with the servant."

"Mé, I want to go, too!" Irene cried. "I can ride well. You yourself said so."

"Mé, Mé!" wailed little Eve, staggering over to her mother. She did not know what was about to happen. But if Irene wanted to do it, she wanted to do it, too.

Marie gathered the two children into her arms. "We shall not be gone long," she told them gaily. "And when we return, we'll all go to the farmer's house for the milk," she promised them.

"And you shall ride on my shoulder, Eve," her father said.

"Good!" Irene cried. "And may I take my butter-fly net?"

At the age of eight, Irene had become greatly interested in collecting butterflies.

"Of course you may," her mother said. "But while Papa and I are gone, you must look after your baby sister."

"I will, Mé," Irene promised.

Churchbells were ringing in the distance as Marie and Pierre pedaled lazily along a country road. Marie looked almost like a young peasant woman in her loose white blouse and wide skirt. Pierre, at her side, was dressed in a dark shirt, ragged trousers, and a battered old cap. For a while they rode along in silence. Then Marie exclaimed, "Pierre, look at all those buttercups! Let's pick some of them."

They got off their bicycles and went down on their knees in the meadow beyond the road. Soon they had each gathered a big bunch of buttercups. Pierre stared in admiration at Marie, with her blond hair, surrounded by the yellow flowers.

"Life has been wonderful with you, Marie," he said softly.

The next evening, Pierre returned to Paris, as he had some work to do. Marie stayed at St. Rémy with Irene and Eve until Wednesday evening. Then they, too, returned home. It was just as well, because the spring weather had changed suddenly. A cold wind blew, and rain fell in torrents.

Thursday morning, Pierre called from downstairs, "Marie, are you going to the lab today?"

"I don't think I shall have time," Marie called back. She was upstairs dressing the children. Irene was now going to a small school some distance from the Rue Kellermann, and already it was late. After Marie took Irene to school, she had errands to do and some shopping. Little Eve would stay home with her grandfather and the servant.

It was six o'clock before Marie returned home from her shopping trip. She was humming contentedly as she put the key in the lock. She had been able to find almost everything she had wanted, and at reasonable prices, too.

The moment Marie stepped inside the door, she knew that something was terribly wrong. Old Dr. Curie stood in the hallway. With him were two old friends, Dr. Paul Appell and Professor Jean Perrin. The look on their faces filled Marie with dread.

Dr. Appell approached and stood before her.

"Marie," he said in a low voice, "there has been an accident. Pierre—he was crossing the Rue Dauphine. A big truck drawn by two horses was coming. He slipped on the wet pavement and— Marie, Pierre is dead."

Marie's face went gray. She stood silent for so long that Dr. Appell feared he would have to repeat his words. Finally she whispered:

"Pierre is dead? Dead? Absolutely dead?"

Professor Perrin came to her side. "What can I do, Marie?" he asked. "Do you wish Madame Perrin to take the children for a while?"

Marie stared at him blankly. When she finally spoke, her voice was hardly above a whisper.

"Irene, yes, please, if you will for a few days." *Pierre dead! It could not be true. Why did not someone say it was not true?* "And please, Jean, send a telegram to my brother in Warsaw. Say— 'Pierre dead result accident.' "

The queer, dry little voice stopped speaking. Marie wandered past the three men and went out through a French door to the dripping garden. She sat down on a wet marble bench and put her face in her hands.

Old Dr. Curie, his own face streaked with tears, frowned worriedly at the lonely little figure. Marie seemed like a body without a soul.

A few days after Pierre's burial at the family tomb in Sceaux, an official of the French government called. He told Jacques Curie, Pierre's brother, that the government wished to give Madame Curie and her children money to live on for the rest of their lives.

When Jacques told Marie, she said, "I don't want a pension. I am young enough to earn my living and that of my children."

But Jacques, and Joseph and Bronya, who had come to be with Marie, were worried. What would become of her? Finally they suggested to the heads of the Sorbonne that Marie take her husband's place as Professor of Physics.

Actually, there was no one else who could have taken Pierre's place. And although up to that time, no woman had ever been given such a position at the university, the heads of the Sorbonne agreed.

When old Dr. Curie told Marie she was being offered the position which Pierre had held, and that
it paid a salary of 10,000 francs a year, she was
quiet for a moment. Perhaps it would make life
more bearable if she went on with Pierre's work.
But perhaps it would be foolish even to think of doing it.

Into her mind came some words Pierre had once
said:

*"Whatever happens . . . one must work just
the same."*

Marie looked up at her father-in-law.

"I will try," she said.

The course she was to teach would begin in November. Marie decided to stay in Paris all summer
and prepare the notes for her lectures. She would
also study Pierre's notebooks, which had ended
abruptly with the words: "When one considers the
progress that has been made in physics . . ."

"I do not wish Irene and Eve to pass the summer
here in the heat," she told Dr. Curie. "Would you
be willing to take Eve with you to St. Rémy?"

"Of course, Marie," the old doctor said kindly.
"But what about Irene?"

"My sister Hela will be at the seashore this summer," Marie said. "She would take Irene, I know. And Irene would love the beach." She paused and looked at her father-in-law. "There is one thing more," she said slowly. "I do not think I can bear to live in this house much longer. I will stay here this summer, but in the fall let us find a house in Sceaux."

Dr. Curie understood why Marie wished to do this. It was hard for her to speak Pierre's name, because she missed him so terribly.

In the fall, Marie moved to Sceaux. All summer long she had bravely tried to prepare herself to carry on Pierre's work. Now the time was coming when she must appear at the Sorbonne.

On Monday, November 5th, at half-past one, the sloping amphitheater of the Hall of Science was filled to overflowing. Students, reporters, the curious public—all tried to jam into the circular tiers of seats. For this was the day Madame Curie was to give her first lecture.

For days, the newspapers had been full of the coming event. Not only because this was the first time a woman had ever lectured at the Sorbonne.

It was also because everyone knew of the two famous scientists and of Pierre Curie's tragic death. An air of excitement filled the hall.

At exactly half-past one, a door at the back opened. Madame Curie entered. Thunderous applause greeted her. Dressed in black, her face pale, she walked to the long table on which stood the apparatus she would need during her lecture. She gazed out over the circle of faces, and for a moment her throat closed up.

Then Marie took a firm hold of the edge of the table. In a calm voice she began to speak:

"When one considers the progress that has been made in physics . . ."

She had started her lecture at the point where Pierre Curie had left off.

CHAPTER FOURTEEN

A Dream Comes True

Mé, I cannot find my turtle. I have looked for him everywhere."

Six-year-old Eve stood anxiously before her mother. The celebrated Madame Curie was sitting on the floor, surrounded by scientific notes. It was Marie's habit to work in this fashion. It made her feel free to spread out her books and papers.

"Where were you playing with your turtle, Eve?" she asked.

"In the garden, Mé," the little girl replied. "The kittens teased him, and he hid from them. But now *I* can't find him either."

Her mother got to her feet and took the child by the hand.

"Come," Marie said. "Show me where you saw him last. He can't have gone far."

Eve led her mother to the far end of the garden and stopped at the edge of one of the narrow paths.

"It was here," she said, pointing. "Oh, my poor turtle!"

"How long ago did you lose him?" Marie asked.

"It must be five minutes ago," Eve said worriedly.

"Then he could not have gone much more than fifteen feet from this spot where you last saw him," her mother said. "Now you must go in a circle about fifteen feet from this spot. And look under the leaves and grass on each side of you."

The little girl dropped to her knees and started creeping in a circle as Marie had suggested. She carefully turned back the thick tufts of grass and looked under the leaves. She had almost completed the circle when, with a shout of triumph, Eve held up the missing turtle. The little fellow waved his flippers feebly, as if to say, "I'm glad you found me."

Marie smiled. "You have just performed an experiment in time and space, little Eve," she said.

"I have, Mé?" Eve said solemnly. "I thought I just found my turtle."

Her mother smiled again. "Come, dear. Let us find Irene. It is almost time for luncheon."

"She is on the trapeze. I saw her before I came in to you," Eve said.

The house which Marie had taken in Sceaux had a large garden. At one end of it she had placed a trapeze, flying rings, and a rope to climb. She wanted Irene and Eve to grow up strong and healthy. For this reason she also went swimming with them whenever she had the time, and took them on bicycle rides.

It was not easy for her to find the time. The train

trip to Paris took half an hour. And often Marie did not return from the university or the laboratory until late in the evening. When she reached home, she would immediately stoke the big coal stove in the hall. Then she would lie down on the couch in the living room for a few minutes to rest.

The children would come and tell her about the happenings of the day. They had a housekeeper-governess who looked after them. But they still missed their grandfather, who had died the year before. Irene, in particular, missed the old doctor, for he had been her constant companion and had taught her many things.

Irene was on the trapeze, as Little Eve had said. When the older girl saw her mother and sister, she swung down and dropped lightly to the ground.

"I knew it must be near lunch time," Irene laughed. "Because I'm so hungry." She gave her mother a quick hug. "I'm so glad it's Sunday," she exclaimed. "We scarcely see you any other day."

Unfortunately, this was only too true. At the Sorbonne, the only course on radioactivity in the world was being given by Madame Curie. The year before, a nine hundred and seventy-one-page book about radioactivity had been published by Madame Curie. A six hundred-page book, "The Works of Pierre Curie," also had been collected and arranged by Madame Curie.

Added to all this, Marie still taught school and continued her work with radium. This work was to win for her the 1911 Nobel Prize in Chemistry at the end of the year.

But one hour at the beginning of each day was reserved for Irene and Eve. At this time, Marie taught them sewing, modeling, arithmetic—not always the same subject, but always something interesting.

Today, as Marie and her daughters sat at luncheon, she was thinking about that morning hour of work.

"You know, Irene," she said, "it is time now that you went to school."

"I know, Mé," Irene sighed. "When I grow up I want to be a scientist just like you. But I hate the thought of being cooped up all day in a stuffy old schoolroom. I'd never have time to work in my garden."

"I don't want you to be indoors all day long,

either," her mother agreed. "I have an idea," she continued, "but I shall have to talk it over with the professors at the university."

"What is your idea, Mé?" little Eve piped up.

"Well," her mother said, "counting your sister Irene, there are about ten children whose parents are professors at the Sorbonne. Why can't each professor teach you ten children one subject a week?

"You would study chemistry one day with Professor Perrin. Another day Professor Mouton would teach you literature. And I would give you a lesson in physics. You would study fewer subjects, perhaps, but I think you might learn more about them."

"And we wouldn't have to spend all day in school!" Irene exclaimed. "Mé, do speak to the professors. I think it is a wonderful idea."

And that is how it happened that for about two years, on certain days, the stately old halls of the Sorbonne rang with children's shouts and laughter.

Every Thursday afternoon Marie Curie taught the children simple physics. For this class the ten lively girls and boys went to a room in the School of Physics in the Rue Cuvier. Here Marie showed them ways of working which she had discovered in her own work.

"You must practice your arithmetic just as my little Eve practices on the piano," she told them. "You must get so that you *never* make a mistake. The secret is in not going too fast."

She taught the children to be neat and orderly in their work.

"Don't tell me you will clean your table *afterward*," she scolded Professor Perrin's son, who had made a complete mess of his lab table. "One must *never* dirty a table during an experiment."

One Thursday afternoon she asked the children, "What would you do to keep the liquid in this jug hot?"

Irene said thoughtfully, "You could wrap the jug in wool."

Professor Langevin's son suggested placing the jug near a hot stove. There were other ideas, none of them very good.

"Well, if I were doing it," Madame Curie said, smiling, "I should start by putting the lid on."

The ten youngsters all laughed. They had not been using common sense. At that moment the door opened. A servant brought in a tray piled high with chocolate bars, sweet buns, and oranges. This meant the lesson was over. The children pounced on the good things. Munching happily, they pushed and crowded each other out the door.

As their voices died away, Marie put away the equipment that they had been using. Then she put on her hat and coat and hurried out of the building. She walked to the site where, at last, the laboratory of Pierre's dreams was being built. This laboratory was to be part of a building devoted to the study of radium. It was to be Marie's laboratory, and she had very definite ideas about it. She wanted it to be large and well lighted. She wanted it to be useful for many years to come—a "temple of the future," as Louis Pasteur, another great scientist, had called it.

Today, when she reached the Rue Pierre Curie, on which the Institute was being built, she went straight to the rambler roses. She had planted them herself earlier, and she watered them every day.

Marie had also had young plane trees planted. They would be fully grown by the time the white walls of the Radium Institute were finished.

It was because of the Radium Institute that Marie refused an invitation to run a new laboratory in Warsaw. In May of 1912, several Polish professors had called on Madame Curie in Paris and begged her to come back to Poland. But that would have meant giving up Pierre's dream of a fine laboratory. She could not do that. So Marie agreed to send her two best assistants to run the Polish laboratory.

Although the Russians still ruled Poland, they made no objection to this. And Marie promised to attend the opening ceremonies when the building was ready.

When the Polish laboratory was finished, Marie journeyed to Warsaw. She was given a wonderful welcome by her countrymen. In the evening she gave a lecture on radium in the Polish language. And she found herself in the same building where, over twenty years before, she had worked in the hidden laboratory of the Floating University.

The next day, Marie was the guest of honor at a banquet. Half hidden behind a large centerpiece of flowers sat a little, white-haired old lady. She gazed at Madame Curie with such admiration that Marie caught her eye.

"It can't be!" Marie thought. "But it is—it's Mademoiselle Sikorska, the directress of the school where the Russian inspector used to pounce on us unexpectedly when we were children."

Impulsive, Marie rose from her place and walked over to where the old lady sat. Leaning over, she kissed her old teacher on both cheeks. This time it was Mademoiselle Sikorska who, like the frightened little girl of long ago, burst into tears.

CHAPTER FIFTEEN

"Little Curies"

I WONDER what I had better do," Marie thought. She was sitting on a high stool in the shining laboratory of the Radium Institute. The recently completed building was very quiet. Not just because it was a Sunday afternoon. But because it was Sunday, August 2, 1914, and terrible things were happening in Europe.

War had broken out between Germany and Russia. Now the Germans were getting ready to declare war on France. Already German soldiers had invaded the country, and soon German hordes would be fighting their way toward its capital, Paris.

Although the people of Paris were trying to remain calm, all the workers in the Institute had left hurriedly to join their regiments. Marie was alone in the building with the cleaning woman and the precious gram of radium that was kept there.

"If I join Irene and Eve at the cottage in the country, I may not be able to get back," she thought. "And if the Germans reach Paris, they will wreck this place. At least, if I am here, they might not touch the lab."

She decided to stay in Paris no matter what happened. But there was one thing that had to be done. The gram of radium in its lead container must be taken to a place of safety.

By September, it began to look as though Paris

would surely be attacked. Marie boarded a train for Bordeaux, far south of Paris. Among the crowds fleeing from Paris, the slender woman in the old black coat was not recognized. Marie carried a small overnight bag in one hand. In the other she carried the heavy case that contained one million francs' worth of radium.

She spent the night in Bordeaux. In the morning she placed the gram of radium in the vault of a bank. Then she took the train that was returning to Paris. It was a troop train, and she was the only person on it who was not a soldier. A little crowd gathered to see this strange woman who was going back to a city which might soon be invaded by the enemy.

When Marie reached Paris, the news was happier. The German advance on the capital had been stopped. A terrible battle at the Marne River was even then taking place.

Irene and Eve had begged to be allowed to come home. Since Paris seemed to be safe now, Marie sent for them. Eve, who was nine, went back to school and her piano lessons. Irene, who was nearly seventeen, started training to become a nurse.

There would be a great need of nurses, for World War I promised to be long and bloody. Many Frenchwomen had volunteered to work in the hospitals. But Marie preferred to help in another way. She was shocked to find that the French hospitals near the battlefields, and many others, too, had almost no X-ray equipment.

These X-ray machines made it possible to "see" instantly where a bullet or a piece of shell had struck a soldier. Hours of suffering could be spared the wounded by the use of X-rays, and many lives saved. But even the ambulances had no X-ray equipment. How could it be rushed to where it was needed?

"In cars, of course," thought Marie. "I must get hold of some cars. We will equip them here at the lab with X-ray machines, and put a dynamo in each car. That will provide electricity for the machines so they can be used on the battlefields and in hospitals at the front. Now what I must do is get the cars."

Before long, Marie had persuaded twenty friends to give her their cars and had equipped them with X-ray machines. The army called them "Little

Curies," and they did a great deal of good, as the war dragged on into another year.

One of the cars, a gray Renault, was kept for Madame Curie's own use. One morning in April, a telegram came to the laboratory. An X-ray car was needed immediately at the hospital in the town of Forges.

Marie told her army driver to fill the gasoline tank. Then she hurried home. By the time the driver pulled up in front of the house in the Quai de Béthune, where Marie and her daughters now lived, she was ready.

Dressed in a big, dark coat, a shapeless old felt hat, and carrying a cracked leather bag, she climbed into the open front seat beside the chauffeur. Soon they were driving at top speed over the rutted roads to Forges. They could hear the sound of the big guns rumbling.

When they reached the hospital, Marie quickly set up the X-ray equipment in one of the rooms. A cable attached to the motor of the Renault outside provided the electric current needed to run the machine.

Marie placed black curtains, which she had brought along over the windows, so the X-ray pictures could be seen clearly on a screen by the surgeon. In half an hour everything was ready for the wounded to be brought in.

The first wounded soldier they examined was hardly more than a boy. He lay on the examination table and stared terrified at the apparatus that loomed over him.

"Will it hurt, what you're going to do?" he asked hoarsely.

"Not at all," Marie assured him in her sweet voice. "It's just like taking a photograph."

She turned on the machine, and on the screen appeared the outline of the boy's hip bone. A dark smudge on the picture showed where a bullet had shattered the bone. The surgeon gave an assistant instructions, and the boy was taken away to await an operation.

It was late when all the wounded soldiers had been X-rayed. But before Marie started back to Paris, she had planned how to set up a permanent X-ray room in this hospital. It would be one of more than two hundred such rooms that she was to set up before the war ended.

On the road back, the driver put on extra speed. Marie, who was riding in the rear seat now, with the X-ray plates, was bounced and jounced against the heavy cases. Somewhere behind them, a bomb fell with a terrific explosion and a flash of glaring light.

Whether the falling bomb upset the driver, or whether the car hit a shell hole in the pitted road, Marie never knew. But the next moment the car turned over and landed upside down in the ditch. Marie was buried under the heavy cases of X-ray equipment and plates.

She could not move under the heavy weight that pinned her down. But her first thought was of the X-ray plates. They were all cracked!

The driver, who had not been hurt, came running. "Madame! Madame!" he was chattering in a frightened voice. "Are you dead?"

"No, I am not dead," Madame snapped. "Come here and pull these cases off me."

It was late when Marie reached home that night. She was sore and bruised but not seriously hurt. She spoke briefly to Irene and Eve, who were at home. Then she went into her own room and shut the door. The two girls looked at each other. What could have happened to upset Mé?

Not until they read a newspaper account of the accident, several days later, did Irene and Eve learn what had "upset" Mé.

After four long years, the First World War was finally won by France and her allies. At eleven o'clock of a Monday morning in November, 1918, came the order for which everyone had been praying: "Cease fire."

All Paris went mad with joy. And Marie Curie was doubly joyful. For not only were the Germans beaten, but her beloved Poland had at last become free of the Russians.

Caught up in the wild enthusiasm of the crowds, Marie and a woman friend hastily decorated the laboratory with French flags. Then they got out the old Renault and drove to a great square in Paris called the Place de la Concorde. There the crowds were so dense that they had to stop. Several persons climbed to the roof of the car and sat there. They cheered loudly when the Renault began to move again.

Victory was sweet, and to help win it Madame Curie had done her bit.

CHAPTER SIXTEEN

A Gift from America

IF YOU had the whole world to choose from, what would you take?"

This fairy-story question was asked of Marie Curie on a May morning in 1920, two years after the war was over. But the person who asked the question was not "making believe."

Her name was Mrs. William Brown Meloney, and she was the editor of an important New York magazine. She had come all the way from America just to meet the famous woman scientist whom she had admired for so long.

Marie and Mrs. Meloney were sitting in the small waiting room of the Radium Institute. Marie's hair was quite gray now, and she wore dark-rimmed spectacles. Her black cotton dress made her pale face seem even paler. Mrs. Meloney thought it was the saddest but most beautiful face she had seen in a long time.

"If I had all the world to choose from?" Marie smiled at her visitor. "Well, I need a gram of radium. But it would cost 500,000 francs. I cannot afford it."

"You, who discovered radium? You have none yourself?" Mrs. Meloney was amazed.

"We have a gram here," Marie said. "I brought it back from Bordeaux during the war, and we used it to help the wounded. But that belongs to the Institute."

Mrs. Meloney said nothing more about radium at that time. But when she returned to the United States, she started a Marie Curie Radium Fund among the women of America. Inside of a year, the necessary $100,000 had been raised.

Soon Marie Curie received an invitation to come to the United States and accept her gram of radium from the President himself.

Marie felt very grateful for the wonderful gift. But she was terrified at the thought of the long journey and of being the center of attention. Mrs. Meloney promised to make the trip as easy as possible, and invited Eve and Irene to come too. At last Marie consented.

Irene and Eve, excited and happy, helped their mother buy some new dresses and a large black hat. The trunk was packed. The day came when they were to board the S.S. *Olympic* bound for America. It was a morning in May, just a year since Mrs. Meloney's first visit. The American lady had come over to Europe to travel back with Marie and the girls.

The steward led them to the most luxurious suite on the ship. Marie's cabin was filled with flowers sent by well-wishers. As the *Olympic* sailed across the ocean, everybody did his best to make Madame Curie comfortable. But Marie was glad when the end of the crossing was near. The sea had been rough and the skies gray almost all the way over. And the motion of the ship made her feel weak and dizzy.

At last the towers of New York appeared on the shimmering horizon. Marie and her two daughters stood on the boat deck as the *Olympic* steamed majestically into the harbor.

"Look!" cried Eve, now sixteen years old and beautiful. Her blue eyes were sparkling with excitement. "There's the Statue of Liberty!"

"Mé, look!" Irene pointed to the huge crowd waiting on the pier. "They're all here to greet *you*."

Her mother gasped in astonishment. Bright flags, American, French, and Polish, whipped in the wind above what seemed to be hundreds, no, thousands of people!

At least three hundred Polish-American women holding red and white roses, waved frantically as the boat approached. Girl Scouts with banners were lined up on the edge of the pier. Over the water came the sound of a band playing. The huge crowd pressed closer and cheered as the boat docked.

Mrs. Meloney had a bit chair brought on deck for Marie to sit in. Then the reporters and photographers descended upon her like a flock of hungry sea gulls.

"Look this way, please." "Look over here, Madame Curie," they called. It was five hours before the bewildered scientist was able to leave the ship and drive to Mrs. Meloney's apartment.

There, Marie was free to sip a quiet cup of tea. Her American friend told her of all the colleges, universities, and scientific societies that wished to honor her.

"You packed your mother's cap and gown, of course?" Mrs. Meloney asked Eve. "She will need to wear them for the ceremonies at the colleges."

Marie laughed. "I don't have a cap and gown," she confessed. "At the Sorbonne all the men professors are obliged to wear an academic gown. But since I am the only woman professor, I never bothered to get one."

"Oh, but you must have one," Mrs. Meloney exclaimed. "I will call in a tailor at once."

The long black robe with velvet edging and billowing sleeves was quickly made. Marie was asked to try it on.

"It is so hot," she objected. "And this heavy silk makes the radium burns on my fingers smart. Must I wear this thing?"

When told firmly that she must, Marie sighed. "Very well," she said, "for university occasions then. But I cannot wear this mortarboard. It keeps sliding off my head."

After an exciting week of luncheons, dinners, meetings in her honor, and tours of the eastern women's colleges, Marie was nearly exhausted. But America had fallen in love with her, and everybody wanted to see her. And the greatest occasion of all was still to come.

May 20th was the date set for the President of the United States, Warren G. Harding, to present to Madame Curie the precious gram of radium. On a table in the East Room of the Capitol in Washington was a lead-lined box. In this container the radium would be taken back to France.

Actually, the radium was not now in the box. It would have been dangerous to expose the radioactive metal. But the crowd of important people who waited for the ceremony to begin could inspect the box as much as they pleased.

At four o'clock sharp, the big double doors opened. The President's wife led the little procession into the East Room, walking with the French Ambassador. Then came Madame Curie, escorted by the President himself. Irene and Eve followed, and Mrs. Meloney.

There were several speeches, and finally President Harding presented to Madame Curie a tiny golden key. This key would unlock the leaden box in which the radium would be placed.

After Marie thanked the President warmly for America's gift, he escorted her to the Blue Room. Here a chair had been placed for her. Then the guests all lined up to pass in front of her. Eve and Irene stood beside their mother. As Mrs. Harding pronounced the names of the guests, the two girls greeted them in French, Polish, or English, and shook hands. They did this in order to save their mother's strength as much as possible.

More and more, Eve and Irene had to take their mother's place. Marie by now was terribly tired. She would have loved to see all of America, but this was impossible. She did go with Mrs. Meloney and the two girls to the Grand Canyon in Colorado. Here, while Marie snatched a little rest at the hotel, Eve and Irene rode wiry Indian ponies along the edge of the canyon. They exclaimed with delight at the grandeur of the view.

After a wonderful month in the United States, it was time to sail for home. The precious gram of radium sailed with Marie, locked in the purser's safe. Marie's cabin was a mass of flowers. Heaps of letters and telegrams littered the tables.

Irene, standing in the middle of her mother's cabin, laughed. "Mé," she said, "it will take days for the three of us to go through all those messages. *Now* will you believe that Madame Curie is a famous person?"

Her mother smiled. "I do not wish to be famous," she said. "In science, we must be interested in things, not persons."

CHAPTER SEVENTEEN

The Rose Vine

MADAME, I beg you. Please to eat something.
You did not stop for luncheon, and it is almost three
o'clock now."

The slightly stooped figure of Georges Boiteux,

gardener and handyman, stood pleadingly before Madame Curie. He was holding a tray on which were slices of buttered bread and a bunch of large purple grapes.

Marie was sitting on a bench under the plane trees, in the garden of the Radium Institute. Beside her was a slide rule and some graph paper on which she had been busily figuring for over an hour. She looked up absent-mindedly. "What? Oh, thank you, Georges. Just put the tray beside me here on the bench, please."

Georges placed the tray on the bench. Then he stood with stubborn devotion, waiting. Madame looked up again, and then she laughed.

"Oh, very well, Georges, I shall eat it now," she said, and plucked a purple grape. Georges smiled with satisfaction and started down the garden path.

Marie looked after him affectionately. Everyone in the Institute, Georges, the young students, Marie's assistants—they all tried to take care of her.

Shortly after the trip to America, Marie's overworked eyes had begun to fail. The doctor had said that the lenses in her eyes were growing foggy. She would go blind unless the cataracts were removed.

Marie did not want her co-workers to know that she was unable to see clearly.

She went to all kinds of trouble to keep them from finding out. She would cleverly question someone about what was written in a report, until the person would tell her without realizing it. She wrote the notes for her lectures in large black letters. She put colored signs on her lab equipment.

In spite of all this, her friends at the Institute guessed what was wrong. To save her from unhappiness, they too used all kinds of little tricks to keep Marie from knowing that they knew. Not until her eyes had been operated on successfully did she learn that a double game had been going on.

But her family and friends still worried about Marie. Though she was almost sixty-seven years old now, she still worked from twelve to fourteen hours a day. Her health was failing rapidly.

After the trip to America, she had taken many other trips. To Holland, Belgium, Spain, South America, Czechoslovakia, and Italy. She had never learned to like fame. But she had learned that the name of Curie had the power to help worthy scientific causes. She never refused that help.

Two years earlier, Marie and her sister Bronya had opened an Institute of Radium in Warsaw. Since that time, Marie somehow had found the time to write a long book on radioactivity. But dearest to her heart was the future of science.

Once she had planted her cherished rose vines for the time when the walls of the Institute would be finished. Now Marie planned ahead to help the young scientists of the future.

She raised money for scholarships. She thought of a way to make scientific words and symbols the same all over the world. By this plan, young workers would be able to find out all that was already known about what they were working on, no matter what country they came from. Then they could use their own time to find out something new about their subject.

Most of all, Marie gave her time and knowledge to the students who worked at the Radium Institute. She was strict with them, and they trembled with nervousness when they handed her their reports. But if she was pleased with their work, they were in heaven.

Young Fournier, one of Marie's best students,

came walking along the path now. He stopped in front of her.

"Madame," he said. "If you can possibly spare the time, I have something interesting to show you."

The young man did have an interesting experiment to show Madame Curie. He also wanted to get her to come inside the building. Although it was a sunny day in May, a cold wind had risen.

"You tried the new method, Monsieur Fournier?" Marie asked eagerly. She began to gather up her papers. "Yes, I will come."

As Marie rose from the bench, a wave of dizziness swept over her. She felt so weak that she might have fallen if young Fournier had not steadied her.

"Madame!" he exclaimed in alarm. "You are ill?"

"It is nothing," she assured him. "I'm tired, that's all. Come. Let us go to the laboratory."

After Marie had inspected the young man's experiment, she walked slowly to her own workroom. She tried to set up the familiar instruments with which she was now experimenting on actinium, another radioactive metal. But Marie was more than just tired. She was gripped with chills and fever. She sat down on the lab stool and looked about the room.

"Perhaps I should stop working," she thought. "Take a rest and do a little gardening. But I don't know whether I could live without the laboratory."

It seemed a fearful thought, and Marie tried once again to adjust her instruments. Her hands were shaking, and she had to give up.

She glanced at the clock on the wall. It was half-past three. Just the hour when she sometimes went to the park to see her little granddaughter, Hélène. Irene, who had become a scientist, had married one of the most promising of all the students at the Institute.

But today Marie did not have the strength to visit little Hélène. She called young Fournier and asked him to have her car brought around.

"I have a fever," she told him, "and I must go home."

While waiting for the car, Marie wandered out into the garden again. The hyacinths were coming along well, and the purple irises stood up bravely. Suddenly Marie stopped. One of the rambler rose vines had drooped, and its leaves had lost their fresh green.

"Georges," she called. The little gardener appeared from behind a bush he had been pruning. "Look at this rose vine! You must see to it right away!"

"*Oui*, Madame," he promised. "I will take care of it at once."

"The car is here," young Fournier said, coming up the garden path. "Allow me to give you my arm, Madame."

Marie took the young student's arm, and they walked to the car. Before climbing in, she turned.

"Don't forget, Georges," she called. "The rose vine."

Two months later, young Fournier stood with tears in his eyes before the untouched apparatus in Marie's workroom. To the other students who were with him, he said sadly, "We have lost everything."

Georges had saved the rose vine. But nothing could save Marie. On July 4th, Madame Curie had died. Her death had been caused by long contact with powerful rays. Rays of the priceless element she had disclosed to the world—radium.